INVENTION

19th-century
brace and bit

Cross-bar
wheel

Radio valve

Early Italian
microscope

"Candlestick"
telephone

19th-century
fountain
pens

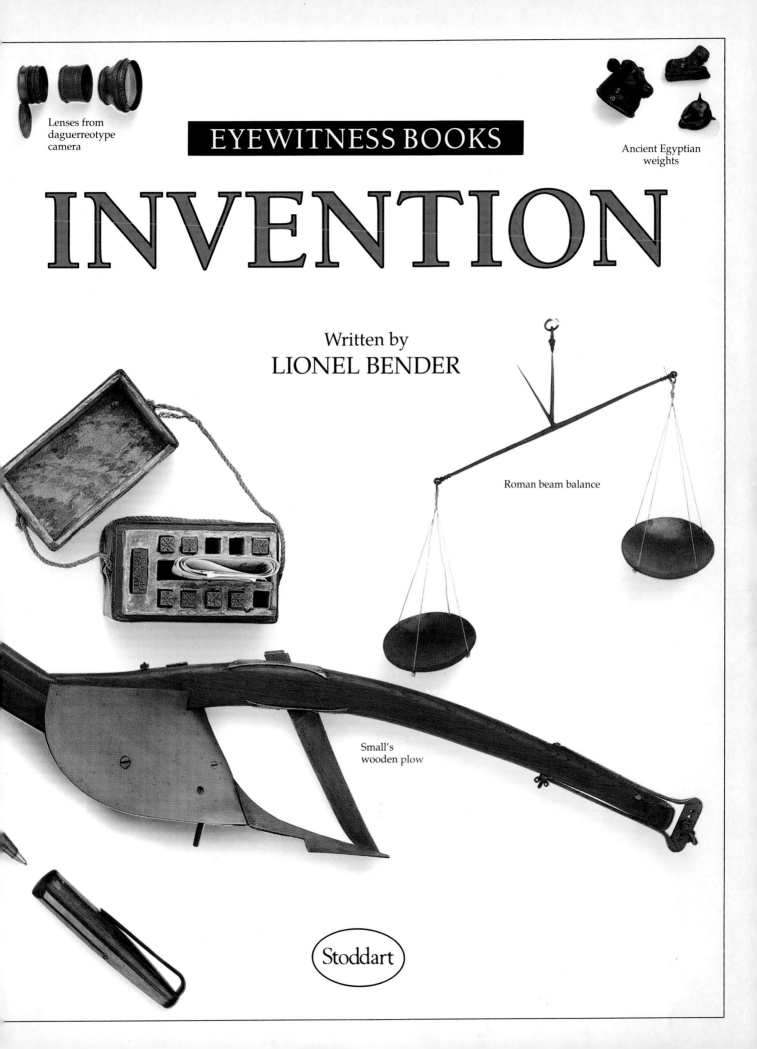

Lenses from
daguerreotype
camera

Ancient Egyptian
weights

EYEWITNESS BOOKS

INVENTION

Written by
LIONEL BENDER

Roman beam balance

Small's
wooden plow

Stoddart

Ivory
portable
sundial

Chinese
measuring
calipers

DK

A DORLING KINDERSLEY BOOK

Project editor Phil Wilkinson
Design Mathewson Bull
Senior editor Helen Parker
Senior art editors Jacquie Gulliver, Julia Harris
Production Louise Barratt
Picture research Kathy Lockley
Special photography Dave King
Additional text Peter Lafferty
Editorial consultants
The staff of the Science Museum, London

19th-century
syringes

Ashanti
gold
weights

This Eyewitness Book has been
conceived by Dorling Kindersley Limited
and Editions Gallimard

Published in Canada in 1991 by
Stoddart Publishing Co. Limited
34 Lesmill Road, Toronto, Canada
M3B 2T6

Published in Great Britain in 1991
by Dorling Kindersley Limited,
9 Henrietta Street, London WC2E 8PS

Canadian Cataloguing in Publication Data

Bender, Lionel
Invention

(Eyewitness books)
ISBN 0-7737-2464-8

1. Inventions - Juvenile literature.   I. King, Dave,
II. Title.   III. Series
T48.B45 1991   j609   C90-095362-4

Stone-headed ax
from Australia

Early
telephone
handset

Colour reproduction by Colourscan, Singapore
Typeset by Windsorgraphics, Ringwood, Hampshire
Printed and bound in Singapore by Toppan Printing Co. (S) Pte Ltd.

Medieval
tally sticks

Contents

Chinese mariner's compass

18th-century English compass

6
What is an invention?
8
The story of an invention
10
Tools
12
The wheel
14
Metalworking
16
Weights and measures
18
Pen and ink
20
Lighting
22
Timekeeping
24
Harnessing power
26
Printing
28
Optical inventions
30
Calculating
32
The steam engine
34
Navigation and surveying
36
Spinning and weaving

38
Batteries
40
Photography
42
Medical inventions
44
The telephone
46
Recording
48
The internal combustion engine
50
Cinema
52
Radio
54
Inventions in the home
56
The cathode ray tube
58
Flight
60
Plastics
62
The silicon chip
64
Index

What is an invention?

AN INVENTION is something that is devised by human effort and that did not exist before. A discovery on the other hand, is something that existed but was not yet known. Inventions rarely appear out of the blue. They usually result from the bringing together of existing technologies in a new way – in response to some specific human need, or as a result of the inventor's desire to do something more quickly or efficiently, or even by accident. An invention can be the result of an individual's work, but is just as likely to come from the work of a team. Similar inventions have even appeared independently of each other at the same time in different parts of the world.

Short handle

Pivot

Arms allow user to adjust depth and direction of cut

Glass bottles

Glass beads

Long blade

Handle

FOOD FOR THOUGHT
The first tin cans had to be opened by hammer and chisel. In 1855, a British inventor developed this claw type of can opener. The blade cut around the rim of the tin using a see-saw levering action of the handle. Openers were given away with a brand of beef, hence the bull's head design.

GLASS
Nobody knows when the process of glass-making (heating together soda ash and sand) was first discovered, although the Egyptians were making glazed beads by 4000 B.C. In the 1st century B.C. the Syrians probably introduced glass blowing, producing objects of many different shapes.

CUTTING EDGE
Scissors were invented more than 3,000 years ago in various places at about the same time. Early scissors resemble tongs with a spring which pushed the blades apart. Modern scissors use the principle of the pivot and the lever to increase the cutting power.

Lid

Bull's head

Blade

IN THE CAN
The technique of heating food to a high temperature to kill harmful bacteria, then sealing it in airtight containers so that it can be stored for long periods, was first perfected by Nicholas Appert in France in 1810. Appert used glass jars sealed with cork, but in 1811 two Englishmen, Donkin and Hall, introduced the use of tin vacuum cans and set up the first food-canning factory.

Lock
mechanism

Iron key

LOCKED UP
In the earliest known locks, the key
was used to raise pins or tumblers
so that a bolt could be moved. Today,
the two most common types are called
the mortise and the Yale.

ZIP UP
The zipper was invented by
engineer Whitcomb Judson in 1891.
It consisted of rows of hooks and
eyes which were locked together
by pulling a slide. The modern
version, with interlocking metal
teeth and slide, was developed
from this early model by
Gideon Sundback and
patented in 1914.

FIRELIGHTERS
Modern matches were
invented by British
chemist John Walker in
1827. He used splinters
of wood tipped with a
mixture of chemicals.
These chemicals were
ignited by heat generated
from the friction of rubbing
the tip on sandpaper.
Matches like this were later
known as lucifers, from the
Latin for "light bearer."

*Bulb from
which air is
extracted*

Sandpaper

PENCIL IN THE DETAILS
Pencil "lead" was invented
independently in France and
Austria in the 1790s. Pencil
makers soon discovered that by varying the relative
amounts of the two main components of the lead
(graphite and clay), they could make leads of different
hardnesses.

*Winder to take up
tape into container*

MASHED UP *below*
Paper was first produced by T'sai Lun in China
in c. A.D 95. The earliest examples were made
from a mixture of cloth, wood, and straw (p. 19).

Metal filament

LIGHTING-UP TIME *left*
The electric light bulb evolved from
early experiments that showed that
an electric current flowing through a
wire creates heat due to resistance in
the wire. If the current is strong
enough, the wire glows white-hot.
There were several independent
inventors, including Thomas Edison
and Joseph Swan. Carbon-filament
lamps were mass-produced from
the early 1880s.

*Circuit
connector*

Paper scroll

**GETTING THE
MEASURE OF IT** *above*
The tape measure evolved from the measuring chains
and rods first used by the Egyptians and then the
Greeks and Romans. This example
incorporates a notebook and
dates from 1846.

Linen tape

*Colter to cut
loose the soil*

IN THE SOIL
The plow developed in about
2000 B.C. from simple hoes and digging
sticks that had been used by farmers for
thousands of years. By changing the shape and size
of its various parts, it was gradually found that
the soil could be cut,
loosened, and turned
in one operation.

*Harness link to
attach team of horses
or oxen*

*Share to cut loose
top layer of soil*

*Moldboard to lift
and turn soil*

The story of an invention

THE CREATION OF AN INVENTION often involves many people, and inventions can take a long time to reach their final form. Sometimes an invention can take centuries to evolve, as the effects of different developments and new technologies are absorbed. After tracing the history of drilling tools, it is apparent that the invention of the familiar hand drill and bit evolved from refinements to the simple awl and the bow drill, over hundreds of years. Among the earliest tools for boring holes were those used by the ancient Egyptians. Around 230 B.C. the Greek scientist Archimedes explored the use of levers and gears to transmit and increase forces. But it was not until the Middle Ages that the brace was developed for extra leverage; the wheelbrace drill, which uses gears, evolved even more recently.

The position in which a bow drill was used

Wooden bow

Cord

Mouthpiece

Bone bow

Wooden handle

Leather strip

Wooden hearth

Metal point

37
7-14
168

HOLE IN ONE
The ancient Egyptians used this early awl to make starter holes for a bow drill bit and to mark the points on planks where wooden pegs were to be fitted.

HOT SPOTS
We do not know whether the bow drill was first developed for woodworking or fire making. The example above is a fire drill. With a bone as the bow, a leather strip was used to rapidly rotate a wooden drill on a wooden hearth. Friction between the drill and the hearth generated enough heat to ignite some dry straw. It also made a hole in the wood.

BORING AWAY
Metal or flint bits were fitted to the drill shaft. A heavy pebble could be used to push down on the shaft to apply more pressure to the bit.

Metal bit

GET THE POINT
A combination of the awl and the simple bow drill produced this Egyptian drill with a metal bit. Various different bits could be used to make wider or narrower holes.

SCREWED UP
Archimedes invented the screw pump. It was a product of his understanding of the inclined plane – essentially it was a rolled-up inclined plane and was used to raise water. The principle of the screw was not used in drill bits until much later.

Main handle

Winch

Pinion

Wider thread allows waste to be removed

Screw thread

FULL CIRCLE
The gimlet has a threaded tip. It can be worked deeper and can make wider holes than a bradawl, with little effort. It is used to make starter holes for screws. The handle is rotated, clockwise to work the tool in, counterclockwise to remove it.

BRACE AND BIT
Bow drills could not transmit enough turning force to drill a wide diameter hole or to drill into tough materials. Using a knowledge of levers, the brace was developed as a means of increasing the turning force. The cranked handle provided leverage. The wider the sweep, the greater the leverage you could obtain.

Grip

Main wheel

Pinion

AUGER
Corkscrew-like bits, or augers, used with a brace, have side grooves that remove waste wood from the hole as the bit bites in. Screwdriver bits can be used with a brace, which provides more turning force than is possible with an ordinary screwdriver.

Mechanism to secure drill head

Chuck

Selection of bits

WHEELBRACE
With gears, the brace drill was adapted for working in more confined spaces and for easy control. Gears were added to transmit the turning force at the handle. With about 80 teeth on the main gearwheel and 20 on the pinions, the bit is rotated 4 times for each turn of the main wheel.

Screwdriver bit

Auger

Chuck

Tools

ABOUT 3.75 MILLION YEARS AGO our distant ancestors evolved an upright stance, and began to live on open grassland. With their hands free for new uses, they scavenged abandoned carcasses and gathered plant food. Gradually, early people developed the use of tools. They used pebbles and stones to cut meat and to smash open bones for marrow. Later, they chipped away at the edges of the stones, so that they would cut better. By 400,000 years ago, flint was being shaped into axes and arrowheads, and bones were used as clubs and hammers. About 250,000 years ago, humans discovered fire. Now able to cook food, our recent ancestors created various tools for hunting wild animals. When they started to farm, a different set of tools was needed.

DUAL-PURPOSE IMPLEMENT
The adze was a variation on the ax that appeared in the 8th millenium B.C. Its blade was set almost at a right angle to the handle. This North Papuan tool could be used either as an ax (as here) or an adze, by changing the position of the blade.

Stone blade

Split wooden handle

STICKY END
This ax from Australia represents the next stage of development from the hand ax. A stone was set in gum in the bend of a flexible strip of wood, and the two halves of the piece of wood were bound together. The ax was probably used to kill wild animals.

GETTING STONED
This flint handax, found in Kent, England, was first roughly shaped with a stone hammer (above), then refined with a bone one. It is perhaps 20,000 years old. It dates from a period known as the Old Stone Age, or Paleolithic period, when flint was the main material used to make tools.

WELL-BRONZED
The use of bronze for tools and weapons began in Asia about 8,000 years ago; in Europe the Bronze Age lasted from about 2000 to 500 B.C.

NEXT BEST THING
Where flint was not available, softer stones were used for tools, as with this rough-stone axhead. Not all stones could be made as sharp as flint.

AX TO GRIND
To make this axhead, a lump of stone was probably rubbed against rocks and ground with pebbles until it was smooth and polished.

HOT TIP
Bow drills were first used to rub one stick against another to make fire. The user moved the bow with one hand and held the shaft steady with the other.

BLOCKHEADS
To drill holes in stone construction blocks, like this practice piece, some early peoples used flint drill heads. These were probably attached to the ends of forked sticks which the masons rotated rapidly by rubbing them between their hands.

A CUT ABOVE THE REST
The ancient Egyptians, probably the most successful of the early civilizations, used stone tools at first. Later they made tools and weapons in ivory, quartz, copper, bronze, and, around 1000 B.C., iron. They also developed wooden rulers and squares.

ALL STRUNG UP
This recent pump drill from New Guinea was fitted with a cast-iron bit. It was used to drill holes in wood. The bowstring, twisted around and secured to the shaft, makes the shaft turn as the stick is pumped downward.

Hole drilled by flint

Bow of twine

Wooden crosspiece

Flint drilling tool

CHISELING AWAY *below left*
In the Stone Age, stone tools, such as this early Danish chisel, or gouge (left), were ground and polished using other rock materials. In ancient Egypt, bronze chisels (center) and chisel blades (right) fitted to wooden handles were used to cut mortise and tenon (interlocking) joints when making wood furniture.

Stone weight

HICK, HACK, HOCK
This adze from Fiji has a handle with a backward-pointing blade providing a good cutting edge. The blade is thick in cross section, so the tool was probably used for heavy-duty work, perhaps hollowing out tree trunks to make boats.

Cord to secure blade

Stone chisel   Bronze chisel

Sharpening stone

Stone blade

Stone tip

SHARP AS A KNIFE
The ancient Egyptians sharpened their bronze tools, and probably their swords and daggers too, by scraping the cutting edges on a smooth lump of sandstone.

JAGGED EDGE *right*
Woodworking as a craft began in Egypt around 3000 B.C. Egyptian carpenters made fine objects to be buried in the tombs of the pharaohs. This cast from an early flint knife, chipped to form a series of teeth, represents one of the earliest examples of a saw.

MIND YOUR TOES
An adze could be used to hack at wood by holding it up in front of your head then swinging it down hard between your legs.

Serrated (notched) edge

The wheel

THE WHEEL is probably the most important mechanical invention of all time. Wheels are found in most machines, in clocks, windmills, and steam engines, as well as in vehicles such as the automobile and the bicycle. The wheel first appeared in Mesopotamia, part of modern Iraq, over 5,000 years ago. It was used by potters to help work their clay, and at around the same time wheels were fitted to carts, transforming transportation and making it possible to move heavy materials and bulky objects with relative ease. These early wheels were solid, cut from sections of wooden planks which were fastened together. Spoked wheels appeared later, beginning around 2000 B.C. They were lighter and were used for chariots. Bearings, which enabled the wheel to turn more easily, appeared around 100 B.C.

POTTER'S WHEEL
By 300 B.C. the Greeks and Egyptians had invented the kick wheel. The disk's heavy weight meant that it turned at near constant speed.

Tripartite wheel

Protective shield for driver

Fixed wooden axle

Peg to hold wheel in place

STONE-AGE BUILDERS *left*
Before the wheel, rollers made from tree trunks were probably used to push objects such as huge building stones into place. The tree trunks had the same effect as wheels, but a lot of effort was needed to put the rollers in place and keep the load balanced.

Solid wooden surface

Axle

Axle

Wooden cross-piece

Axle

SCARCE BUT SOLID
Some early wheels were solid disks of wood cut from tree trunks. These were not common, since the wheel originated in places where trees were scarce. Solid wooden chariot wheels have been found in Denmark.

PLANK WHEEL
Tripartite or three-part wheels were made of planks fastened together by wooden or metal cross-pieces. They are one of the earliest types of wheel and are still used in some countries because they are suitable for bad roads.

ROLLING STONE
In some places, where wood was scarce, stone was used for wheels instead. It was heavy, but long-lasting. The stone wheel originated in China and Turkey.

WHEELS AT WAR
The wheel made possible the chariot, which originated in Mesopotamia around 2000 B.C.

Leather thongs

Wooden chassis beam

LEATHER BEARING
Around 100 B.C., the Celts of France and Germany made carts with simple axle bearings. These consisted of leather sleeves that fitted between the axle and the wheel hub. They reduced friction, allowing the wheel to turn easily.

CROSSBAR
The horse was strapped to the crossbar, which was bound to the chassis with leather thongs.

Peg to hold wheel in place

Chassis

Fixed axle

FIXED IN PLACE
The fixed axle was rigid. It was attached to the chassis of the vehicle. The wheel turned around the axle.

Wheel

HARVEST
Wheels like this, with metal rims to lessen wear, were made as early as 2000 B.C. They were used throughout the middle ages.

Chassis

Rotating axle

MOVING AXLE
The moving axle was fixed rigidly to the wheel and turned with it.

Wheel

Roller bearings

ROLLERS
Around 100 B.C. Danish wagon-makers probably tried putting wooden rollers around the axle in an attempt to make the wheel turn more smoothly.

Roller bearings

Early Middle-Eastern cart

Holes make wheel lighter

Axle

Cut stone construction

EARLY SEMISOLID WHEEL
Wheels could be made lighter by cutting out sections of wood. Wheels of this type, called Dystrop wheels, were made around 2000 B.C.

Axle

CROSSBAR WHEEL
If large sections of a wheel were cut away, the wheel could be strengthened with struts or crossbars. From here it was a small step to the spoked wheel.

Spokes to strengthen wheel

Metalworking

GOLD AND SILVER occur naturally in their metallic state. From early times, people found lumps of these metals and used them for simple ornaments. But the first useful metal to be worked was copper, which had to be extracted from rocks, or ores, by being heated on a fierce fire. Then bronze, an alloy, was created by mixing copper and tin together. Bronze was strong and did not rust or decay. It was easy to work by melting and pouring into a shaped mold, a process called casting. Because bronze was strong and easy to work, everything from swords to jewelry was made of the metal. Iron was first used around 1500 B.C. Iron ores were burned in a fire with charcoal, producing an impure form of the metal. Iron was plentiful, but difficult to melt; at first it had to be worked by hammering rather than casting.

Roman iron nail, about A.D. 88

CASTING – FINAL STAGE
When cold, the mold was broken open and the object removed. Solid bronze is far harder than copper and can be hammered to give it a sharp cutting edge. Because of this, bronze became the first metal to be widely used.

Bloom of iron

Iron ore

Partially hammered bloom

BLOOM OF IRON
Early furnaces were not hot enough to melt iron and so the metal was produced as a spongy lump, called a bloom. The bloom was hammered into shape while white hot.

CASTING – FIRST STAGE
The first stage in producing bronze was to heat copper and tin ores in a large bowl or a simple furnace. Bronze is easier to cast into a variety of shapes than copper.

CASTING – SECOND STAGE
The molten bronze was poured into a mold and allowed to cool and solidify. This process is called casting. Knowledge of bronze casting reached Europe by about 3000 B.C., and China several centuries later.

IRON SWORD-MAKING
In the first century A.D., iron swords were made by twisting and hammering together several strips or rods of iron. This process was called pattern welding.

14

PINS AND NEEDLES
Bronze could be worked into delicate, small objects, such as pins and needles. It was also used for large objects such as bells and statues.

ROMAN NAILS
These iron nails were removed from Roman sites in London and Scotland.

The horse hobble, made of wrought iron, was an early form of horseshoe. It was strapped in place over the hoof.

WROUGHT OR CAST ?
Wrought iron is a pure form of iron made in a simple furnace as a pasty lump, which has to be hammered into shape. It was not possible to make molten iron, which could be cast, until after the introduction of the blast furnace in the 1300s A.D.

Loop to accept strap

Flat surface to take base of horse's foot

AFRICAN IRON
Making iron using simple furnaces was still in practice in parts of Africa in the 1930s. These items made in the Sudan were produced in a clay furnace and hammered into shape.

Quer – type of hoe made of wrought iron

Barbed point

GETTING THE POINT
Iron was often used for weapons, which could be quite elaborate. This spearhead had a wooden handle.

BRONZE ORNAMENTS
Bronze bracelets were often decorated with fine patterns. Ornamental hairpins sometimes had large hollow heads covered with patterns.

Bracelet   Hairpin

IRON HAMMER *right*
Iron has been used for hammers for centuries. This simple iron hammer comes from the Sudan and dates from about 1930.

Iron strands bound together for strength

Point made of pieces of iron hammered together

DECORATIVE SWORDS
Pattern welding produced a strong blade that could be sharpened to make a fine, strong cutting edge. The twisted iron strips forming the blade produced an ornamental pattern along its length.

Finished sword

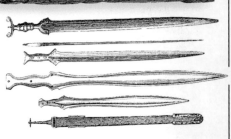

SMALL HANDS? *above*
Bronze swords often had ornamental handles and finger guards. The handles were often very short and could not be held comfortably by hands as large as ours.

Weights and measures

THE FIRST SYSTEMS of weights and measures were developed in ancient Egypt and Babylon. They were needed to weigh crops, measure farmland, and standardize currency for the buying and selling of goods. Around 3500 B.C. the Egyptians invented scales; they had standard weights and a measurement of length called the cubit, equal to about 21 in (53 cm). The Code of Hammurabi, a document recording the laws of the king of Babylon from 1792 to 1750 B.C., refers to standard weights and different units of weight and length. By Greek and Roman times, scales, balances, and rulers were in everyday use. Present-day systems of weights and measures, the imperial (foot, pound) and metric (meter, gram), were established in the 1300s and 1790s, respectively.

Early Egyptian stone weights

Metal Egyptian weights

HEAVY METAL
Early Egyptians used rocks as standard weights, but around 2000 B.C., as metal-working developed, weights cast in bronze and iron were used.

Hook for object to be weighed

WORTH THEIR WEIGHT IN GOLD
The Ashanti, Africans from a gold-mining region of modern Ghana, rose to power in the 18th century. They made standard weights in the form of gold ornaments.

Fish

Scorpion

Sword

WEIGHING HIM UP
This ancient Egyptian balance is being used in a ceremony called "Weighing the heart," which was supposed to take place after a person's death.

Pointer

OFF BALANCE
This Roman beam balance for weighing coins consists of a bronze rod pivoted at the center. Objects to be weighed were placed on a pan hung from one end of the beam and were balanced against known weights hung from the other end. A pointer at the center of the beam showed when the pans balanced.

Pan

Hollow to take smaller weights

WEIGHTY NEST EGGS
With simple balances, sets of standard weights are used. Large or small weights are put on or taken off until the balance is horizontal. These are French 17th-century nesting weights; one fits just inside the next to make a neat stack.

Scale in inches and centimeters

USING THE STEELYARD *right*
On a steelyard, the weight is moved along the long arm, and the distance from the pivot to the balance-point, read off the scale marked on the arm, gives the object's weight. It had an advantage for traveling merchants in that they did not need to carry a large range of weights.

Scale

Movable weight

ALL HOOKED UP
The steelyard was invented by the Romans around 200 B.C. Unlike a simple balance it had one arm longer than the other. A sack of grain would be hung from the short arm, and a single weight moved along the long arm until it balanced. This example dates from the 17th century.

A BIG STEP *above right*
This British size-stick for measuring people's feet starts with size 1 as a 4.33 inch length and increases in stages of one-third of an inch.

FILLED TO THE BRIM *below*
Liquids must be placed in a container, such as this copper jug used by a distiller, in order to be measured. The volume mark is in the narrow part of the neck, so the right measure can instantly be seen.

Volume mark here

STICKING TO ONE'S PRINCIPLES
The first official standard yard was established by King Edward I of England in 1305. It was an iron bar divided into 3 feet of 12 inches each. This is a 19th-century tailor's yardstick used to measure lengths of cloth. It also has a centimeter scale.

Foot positioned here

Adjustable jaw

GRIPPED TIGHT *above right*
Wrenchlike sliding calipers, used to measure the width of solid objects and building materials such as stone, metal, and wood, were invented at least 2,000 years ago. Measurements are read off a scale on a fixed arm, as on this replica of a caliper from China.

FLEXIBLE FRIEND *left*
Tape measures are used in situations where a ruler is too rigid. Measuring people for clothes is one of the most familiar uses of the tape measure; longer tapes are used for land measurement and other jobs.

GETTING IT RIGHT
One of the most important things about weights and measures is that they should be standardized, so that each unit of measure is always of identical value. These men are testing weights and measures to ensure they are accurate.

NO SHORT MEASURES
This Indian grain measure was used to dispense standard quantities of loose items. A shopkeeper would sell the grain by the measureful rather than weigh different quantities each time.

Pen and ink

WRITTEN RECORDS first became necessary with the development of agriculture in the Fertile Crescent in the Middle East about 7,000 years ago. The Babylonians and ancient Egyptians inscribed stones, bones, and clay tablets with symbols and simple pictures. They used these records to establish land holdings and irrigation rights, to keep records of harvests, and write down tax assessments and accounts. At first, they used flints to write with, then they used the whittled ends of sticks. Around 1300 B.C. the Chinese and Egyptians developed inks made from lampblack – soot from the oil burned in lamps – mixed with water and plant gums. They could make different colored inks from earth pigments such as red ochre. Oil-based inks were developed in the Middle Ages for use in printing (pp. 26-27), but writing inks and lead pencils are modern inventions. More recent developments, such as the fountain pen and the ballpoint, were designed to get the ink on the paper without needing to refill the pen constantly.

LIGHT AS A FEATHER
A quill – the hollow shaft of a feather – was first used as a pen around 500 B.C. Dried and cleaned goose, swan, or turkey feathers were most popular because the thick shaft held the ink and the pen was easy to handle. The tip was shaved to a point with a knife and split slightly to ensure that the ink flowed smoothly.

HEAVY READING
The first writing that we have evidence of is on Mesopotamian clay tablets. Scribes used a wedge-shaped stylus to make marks in the clay while it was wet. The clay dried and left a permanent record. The marks that make up this sort of writing are called cuneiform, meaning wedge-shaped.

A PRESSING POINT
In the 1st millenium B.C. the Egyptians wrote with reeds and rushes, which they cut to form a point. They used the reed pens to apply lampblack to papyrus.

余
牧
戔

Chinese characters

ON PAPYRUS
Ancient Egyptian and Assyrian scribes wrote on papyrus. This was made from pith taken from the stem of the papyrus plant. The pith was removed, arranged in layers, and hammered to make a sheet. The scribe (left) is recording a battle. The papyrus (right) is from ancient Egypt.

STROKE OF GENIUS
The ancient Chinese wrote their characters in ink using brushes of camels' or rats' hairs. Clusters of hairs were glued and bound to the end of a stick. For fine work on silk they used brushes made of just a few hairs glued into the end of a hollow reed. All 10,090 or more Chinese characters are based on just eight basic brushstrokes.

Ink reservoir
for early
ballpoint pen

Fiber tip

SOFTLY DOES IT
Fiber or soft-tipped pens were invented in the 1960s. A stick of absorbent material acts as the ink reservoir. The tip-stalk, embedded in the reservoir, contains narrow channels through which ink flows as soon as the tip touches the paper.

Lever for filling pen

Free-moving ball

ON THE BALL
The ballpoint pen was developed by John H. Loud in the 1880s. The modern version was invented by the Hungarians Josef and Georg Biro in the 1940s. At the tip of an ink-filled plastic tube is a tiny free-moving metal ball. Ink flows from the tube through a narrow gap to the ball, which transfers the ink to the paper.

CLOGGING UP THE WORKS
Fountain pens were invented in Europe around 1800. Rubber tubing, inside a metal stem, was used to hold the ink, which was a solution of natural plant dyes such as indigo. Unless the dyestuff was finely ground, the ink would clog the nib. In 1884, Edson Waterman invented the first true fountain pen.

HIS NIBS
Dip pens, like those used in schools until the 1960s, had a wooden stem, metal nib holder, and changeable nibs. Early pen nibs, like these, were all steel. Modern versions are often tipped with hardwearing metals such as osmium or platinum.

Papermaking
The earliest fragments of paper that have been discovered come from China and date from *c.* A.D. 90. Knowledge of papermaking eventually spread to Europe via the Islamic world. The basic process remained similar to that used in China. Paper was made from wood pulp and rags, which were soaked in water and beaten into a pulp.

FIT FOR A KING
The scribes of the Middle Ages used quill pens to produce their elaborately decorated manuscripts. This example records the coronation of King Henry of Castile in the 15th century. It shows the delicate strokes that were possible with quite simple equipment.

Sharpened point

Range of nibs for dip pens

TRAY BY TRAY *right*
A tray with wire grids was lowered into the pulp, the grid removed, and surplus water shaken off.

HANGING OUT TO DRY
The resulting sheet was taken off the grid and put on a piece of felt before finally being hung up to dry.

MISSED THE POINT
Quill pens were worn down by the constant scraping against the rough paper or parchment and from time to time had to be resharpened. In the 17th century, quill-sharpeners were invented. The worn end of the quill was snipped off neatly.

Lighting

THE FIRST ARTIFICIAL LIGHT came from fire, but fire was dangerous and difficult to carry around. Then, some 20,000 years ago, people realized that they could get light by burning oil, and the first lamps appeared. These lamps were hollowed-out rocks full of animal fat. Lamps with wicks of vegetable fibers were first made in about 1000 B.C. They had a simple channel to hold the wick; later, the wick was held in a spout. Candles appeared about 2,000 years ago. A candle is just a wick surrounded by wax or tallow. When the wick is lit, the flame melts some of the wax or tallow, which burns to give off light. So a candle is really an oil lamp in a more convenient form. Oil lamps and candles were the chief source of artificial light until gas lighting became common in the 19th century; electric lighting took over more recently.

CAVE LIGHT
When early people made fire for cooking and heating, they realized that it also gave off light. So the cooking fire provided the first source of artificial light. From this it was a simple step to making a brushwood torch, so that light could be carried or placed high up in a dark cave.

SHELL-SHAPED *right*
By putting oil in the body and laying a wick in the neck, a shell could be used as a lamp. This one was used in the 19th century, but shell lamps were made centuries before.

Wick

COSTLY CANDLES
The first candles were made over 2,000 years ago. Wax or tallow was poured over a hanging wick and left to cool. Such candles were too expensive for most people.

Container for wax

Wicks

Spout for wick

UP THE SPOUT
Saucerlike pottery lamps have been made for thousands of years. They burned olive oil or seed oil . This one was probably made in Egypt about 2,000 years ago.

COVERED OVER *right*
The Romans made clay lamps with a covered top to keep the oil clean. They sometimes had more than one spout and wick to give a stronger light.

Hole for wick

Wick

HOLLOWED OUT *left*
The most basic form of lamp is a hollowed-out stone. This one came from the Shetland Islands and was used during the last century. But similar examples have been found in the caves at Lascaux, France, dating from about 15,000 years ago.

MOLDS
Candles have been made in molds since the 15th century. They made candle-making easier and were widely used in early American homes and shops.

DRY AS TINDER
Before the introduction of matches, tinder boxes were used to light fires and lamps. A spark was made by striking a flint (the striker) against a piece of metal (the steel). Some dry material (the tinder) in the box would catch fire.

Handle

Steel

Tinder

Lid

Candle holder

Striker

Tinder box

LIGHTS OUT
Cone-shaped snuffers were often used to put out candles. There was no smell and little risk of being burned.

Cover to put out fire

TRIMMING THE WICK
With the appearance of more sophisticated oil lamps, elaborate tools were made to cut the wicks. This wick trimmer clips the wick and flicks the debris into a container.

CANDLE POWER *above*
A single candle produces only a little light – one candle power.

PROTECTOR
Lanterns were used to shield the flame from the wind and to reduce the risk of fire.

ON THE STREETS *above*
This engraving shows the first candle street lamp being lit in Paris in 1667. The lamplighter had to climb a stepladder to reach the lantern.

Handle to raise candle

SWEETNESS AND LIGHT
Another way to make a candle was to use wax collected from a beehive. This could be rolled into a cylinder shape.

TWISTER *left*
This candlestick has a spiral mechanism. The user twists it as the candle burns down, to keep the flame at the same level.

Timekeeping

KEEPING TRACK OF TIME was important to people as soon as they began to cultivate the land. But it was the astronomers of ancient Egypt, some 3,000 years ago, who used the regular movement of the sun through the sky to tell time more accurately. The Egyptian shadow clock was a sundial, indicating time by the position of a shadow falling across markers. Other early devices for telling time depended on the regular burning of a candle, or the flow of water through a small hole. The first mechanical clocks used the regular rocking of a metal rod, called a foliot, to regulate the movement of a hand around a dial. Later clocks use pendulums, which swing back and forth. An escapement ensures that this regular movement is transmitted to the gears, which drive the hands.

BOOK OF HOURS
Medieval books of hours, prayer books with pictures of peasant life in each month, show how important the time of year was to people working on the land. This illustration for the month of March is from *Très Riches Heures* of Jean, duc de Berry.

PLUMB LINE
The ancient Egyptian merkhet was used to observe the movement of certain stars across the sky, allowing the hours of the night to be calculated. This one belonged to an astronomer-priest of about 600 B.C. named Bes.

Folding gnomon

Holes to take pin

COLUMN DIAL
This small ivory sundial has two gnomons (pointers), one for summer, one for winter.

Cover

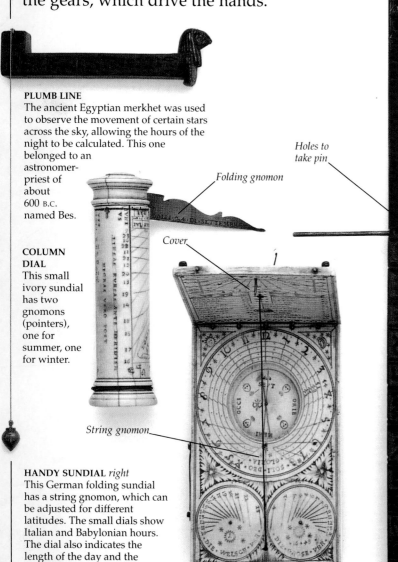

String gnomon

HANDY SUNDIAL *right*
This German folding sundial has a string gnomon, which can be adjusted for different latitudes. The small dials show Italian and Babylonian hours. The dial also indicates the length of the day and the position of the sun in the zodiac.

TIBETAN TIMESTICK
The Tibetan timestick relied on the shadow cast by a pin through an upright rod. The pin would be placed in different positions according to the time of the year.

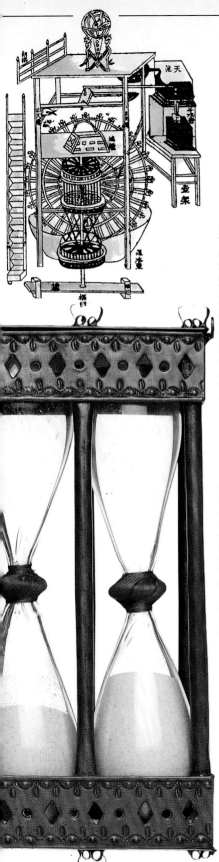

WATER CLOCK
This Chinese water clock was built in 1088. It was housed in a tower 33 ft (10 m) high and relied on the power from a water wheel to work bells, gongs, and drums which marked the hours.

Adjustable weights

LANTERN CLOCK
This Japanese lantern clock was regulated by moving small weights along a balance bar. The clock has only one hand indicating the hour. Minute hands were uncommon before the 1650s, when Dutch scientist Christian Huygens made a more accurate clock regulated by a swinging pendulum.

BRACKET CLOCK *below*
Bracket clocks were first made in the 17th century. This example was made by the famous English clockmaker Thomas Tompion. It has dials to regulate the mechanism and to select striking or silent operation.

CHRISTIAN HUYGENS
This Dutch scientist made the first practical pendulum clock in the mid 17th century.

VERGE WATCH
Until the 15th century, clocks were powered by falling weights, and so could not be moved around. The use of a coiled spring to drive the hands meant that portable clocks and watches could be made, but they were not very accurate. This example is from the 17th century.

BALANCE-SPRING WATCH
Christian Huygens introduced the balance spring in 1675. It allowed much more accurate watch movements to be made. Thomas Tompion, the maker of this watch, introduced the balance spring to England, giving that country a leading position in watchmaking.

SANDS OF TIME *above*
The sandglass was probably first used in the Middle Ages, around A.D. 1300, though this is a much later example. Sand flowed through a narrow hole between two glass bulbs. When all the sand was in the lower bulb, a fixed amount of time had passed.

Harnessing power

SINCE THE DAWN OF HISTORY, people have looked for sources of power to make work easier and more efficient. First they made human muscle power more effective through the use of machines such as cranes and treadmills. It was soon realized that the muscle power of animals, such as horses, mules, and oxen, was much greater than that of humans. Animals were trained to pull heavy loads and work in treadmills. Other useful sources of power came from wind and water. The first sailing ships were made in Egypt about 5,000 years ago. The Romans used water mills for grinding corn during the 1st century B.C. Water is still a valuable source of power and has been in widespread use ever since. Windmills spread westward across Europe in the Middle Ages, when people began to look for a more efficient way of grinding corn.

MUSCLE POWER
Dogs are still used in arctic regions to pull sleds, though elsewhere in the world the horse has been the most common working animal. Horses were also used to turn machinery such as grindstones and pumps.

POST MILL
The earliest windmills were probably post mills. The whole mill could be turned around a central post in order to face into the wind. The mill was made of wood; many mills were fragile and could blow over in a storm.

HAUL AWAY!
This 15th-century crane in Bruges, Belgium, was worked by men walking on a treadmill. It is shown lifting wine kegs. Other simple machines, such as the lever and pulley, were the mainstay of early industry. It is said that around 250 B.C. the Greek scientist Archimedes could move a large ship single-handed by using a system of pulleys. It is not known exactly how he did this.

Tail pole

THE FIRST WATER WHEELS
From around 70 B.C. we have records of the Romans using two types of water wheel to grind corn. In the undershot wheel, the water passes beneath the wheel; in the overshot wheel, the water flows over the top. The latter can be more efficient, using the weight of the water held on the blades.

STANDARD *left*
The Halladay Standard Windmill, introduced in the mid 1800s, is the forerunner of wind pumps still used in remote areas.

ADAPTABLE POWER
In the Middle Ages, water mills were used for tasks from cleaning cloth to blowing bellows for blast furnaces. Later, they were used to drive factory machinery.

Whip

SAILS
Simple sails were made from canvas stretched over a frame. An improved type of sail was invented by Andrew Meikle in the 1770s. It consisted of hinged slats that were kept in place by a spring. When the wind became too strong, the slats opened, allowing the wind to pass harmlessly through.

INSIDE A POST MILL *right*
Inside the mill, the shaft from the sails was attached to a large gear wheel, called the brake wheel. This meshes with another gear, called the stone nut, which was connected to a vertical shaft that turned the runner stone.

Windshaft

Stock

Stone case containing mill stones

Brake wheel with gear teeth to drive runner stone

Sail cloth

Fixed post

SUPPORTING THE MILL
The legs of this post mill are visible, but they were sometimes enclosed in a solid wall. This type of structure evolved into the tower mill, which had a solid tower with a small cap that could be turned to face the wind.

Revolving body or "buck"

TURNING THE MILL
To turn the mill into the wind, the miller pushed the tail pole near the stairs. Later mills had a small wind wheel, called a fantail, with its own small sails. This turned the mill automatically.

Rope to operate sack hoist

Cross trees

Printing

BEFORE PRINTING BEGAN, each copy of every book had to be written out by hand. This made books rare and expensive. The first people to print books were the Chinese and Japanese in the sixth century. Characters and pictures were engraved on wooden, clay, or ivory blocks. When a paper sheet was pressed against the inked block, the characters were printed on the sheet by the raised areas of the engraving. This is known as letterpress printing. The greatest advance in printing was the invention of movable type – single letters on small individual blocks that could be set in lines and reused.

This innovation also began in China, in the 11th century. Movable type was first used in Europe in the 15th century. The most important pioneer was German goldsmith Johannes Gutenberg. He invented typecasting – a method of making large amounts of movable type cheaply and quickly. After Gutenberg's work in the late 1430s, printing with movable type spread quickly across Europe.

This early Japanese wooden printing block has a complete passage of text carved into a single block of wood.

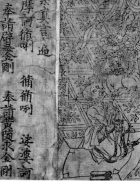

FROM THE ORIENT
This early Chinese book was printed with wooden blocks, each of which bore a single character.

EARLY TYPE
Blocks with one character were first used in China in about 1040. These are casts of early Turkish types.

PUNCHES
Gutenberg used a hard metal punch, carved with a letter. This was hammered into a soft metal to make a mold.

Letter stamped in metal

IN GOOD SHAPE
Each "matrix" bore the impression of a letter or symbol.

POURING HOT METAL
A ladle was used to pour molten metal, a mixture of tin, lead, and antimony, into the mold to form a piece of type.

THE GUTENBERG BIBLE
In 1455 Gutenberg produced the first large printed book, a Bible which is still regarded as a masterpiece of the printer's art.

Screw to secure blade

Metal blade

CLOSE SHAVE
A type plane was used to shave the backs of the metal type to ensure that all the letters were exactly the same height.

TYPE MOLD
The matrix was placed in the bottom of a mold like this. The mold was then closed, and the molten metal was poured in through the top. The sides were opened to release the type.

Mold inserted here

Spring to hold mold closed

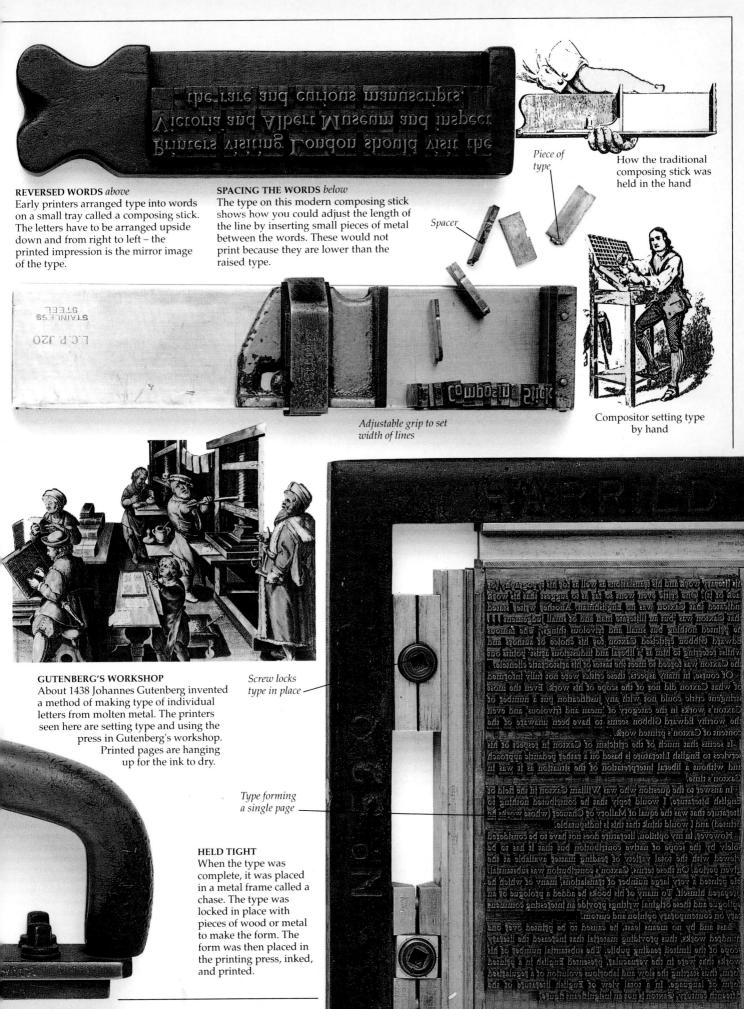

REVERSED WORDS *above*
Early printers arranged type into words on a small tray called a composing stick. The letters have to be arranged upside down and from right to left – the printed impression is the mirror image of the type.

SPACING THE WORDS *below*
The type on this modern composing stick shows how you could adjust the length of the line by inserting small pieces of metal between the words. These would not print because they are lower than the raised type.

Piece of type

Spacer

How the traditional composing stick was held in the hand

Compositor setting type by hand

STAINLESS STEEL

I.C.P. J20

Adjustable grip to set width of lines

GUTENBERG'S WORKSHOP
About 1438 Johannes Gutenberg invented a method of making type of individual letters from molten metal. The printers seen here are setting type and using the press in Gutenberg's workshop. Printed pages are hanging up for the ink to dry.

Screw locks type in place

Type forming a single page

HELD TIGHT
When the type was complete, it was placed in a metal frame called a chase. The type was locked in place with pieces of wood or metal to make the form. The form was then placed in the printing press, inked, and printed.

Optical inventions

THE SCIENCE OF OPTICS is based on the fact that light rays are bent, or refracted, when they pass from one medium to another (for example, from air to glass). The way in which curved pieces of glass, or lenses, refract light was known to the Chinese in the 10th century A.D. In Europe in the 13th and 14th centuries the properties of lenses began to be used for improving vision, and eyeglasses appeared. Later, people used mirrors made of shiny metal as an aid to applying makeup and hairdressing. It was not until the 17th century that more powerful optical instruments, capable of magnifying very small items and bringing distant objects into clearer focus, began to be made. Developments at this time included the telescope, which appeared at the beginning of the century, and the microscope, invented around 1650.

IN THE DISTANCE
The telescope must have been invented many times – whenever someone put two lenses together like this and realized distant objects could be made to look larger.

BLURRED VISION
Eyeglasses, pairs of lenses for correcting sight defects, have been in use for over 700 years. At first they were used only for reading, and like the ones being sold by this early optician, were perched on the nose when needed. Eyeglasses for correcting near-sightedness were first made in the 1450s.

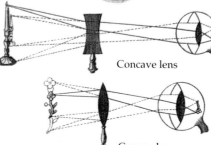

GLASS EYES?
Convex (outward-curving) lenses were known in 10th-century China, but the use of lenses for reading glasses and to make eyeglasses for the far-sighted probably began in Europe. These 17th-century reading glasses use convex lenses.

17th-century eyeglasses

17th-century glass was often colored

Leather-covered tube

Lens cap

STARGAZING
The celebrated Italian scientist and astronomer Galileo Galilei pioneered the use of refracting telescopes to study the heavens. This is a replica of one of Galileo's earliest instruments. It has a convex lens at the front and a concave (inward-curving) lens at the viewing end.

Concave lens

Convex lens

COLORING THE VIEW
Early refracting telescopes, such as this 18th-century English model, produced images with blurred, colored edges, because their lenses bent the different colors of light by different amounts. In 1758, based on earlier work by Chester Moor Hall, Englishman John Dolland developed a main lens free of color distortion by combining lenses of flint (high-color distortion) glass, and crown (low-color distortion) glass.

Eyepiece lens

Objective lens

ANTONI VAN LEEUWENHOEK (1632-1723) *left*
The Dutchman Leeuwenhoek taught himself to grind lenses and made simple microscopes with a tiny lens in a metal frame. Obtaining magnifications of up to 270 times, he was one of the first to study the miniature natural world. He described "very little and odd animalcules" in drops of pond water.

COMPOUND INTEREST *above*
The compound microscope has not one but two lenses. The main lens magnifies the object, and the eyepiece lens enlarges the magnified image.

Lens cap

Lens cap

ON REFLECTION
The reflecting telescope uses a mirror lens. This avoids the problem of color distortion and the need for long focal-length lenses, which required long viewing tubes. This version has two mirrors and an eyepiece lens.

Geared focusing mechanism

Eyepiece

ON THE LEVEL
A quadrant and plumb line are fitted to this 17th-century telescope. They help the astronomer work out the altitude of an object in the sky.

PEEPING TOM
Jealousy glasses were sometimes used by the 18th-century English gentry for keeping an eye on one another. A mirror in the tube reflects the light rays so that you could look to one side when it seemed that you were looking straight ahead.

18th-century pocket telescope

Focus adjuster

TWO FOR A TENOR
Simple binoculars, like these 19th-century opera glasses decorated with mother-of-pearl and enamel, consist of two telescopes mounted side-by-side. Prism binoculars had been invented by 1880. The prism, a wedge of glass, "folded" the light rays, shortening the length of the tube needed and allowing greater magnification in a smaller instrument.

Calculating

PEOPLE HAVE ALWAYS counted and calculated, but calculating became very important when the buying and selling of goods began. Apart from fingers, the first aids to counting and calculating were small pebbles, used to represent the numbers from one to ten. About 5,000 years ago, the Mesopotamians made several straight furrows in the ground into which the pebbles were placed. Simple calculations could be done by moving the pebbles from one furrow to another. Later, in China and Japan, the abacus was used in the same way, with its rows of beads representing hundreds, tens, and units. The next advances did not come until much later, with the invention of calculating aids like logarithms, the slide rule, and basic mechanical calculators in the 17th century A.D.

Upper beads are five times the value of lower beads

USING AN ABACUS
Experienced users can calculate at great speed with an abacus. As a result, this method of calculation has remained popular in China and Japan – even in the age of the electronic calculator.

POCKET CALCULATOR
The ancient Romans used an abacus similar to the Chinese. It had one bead on each rod in the upper part; these beads represented five times the value of the lower beads. This is a replica of a small Roman hand abacus made of brass.

THE ABACUS
In the Chinese abacus, there are five beads on the lower part of a rod, each representing 1, and two beads on the upper part, each representing 5. The user moves the beads to perform calculations.

HARD BARGAIN
Making quick calculations became important in the Middle Ages, when merchants began to trade all around Europe. The merchant in this Flemish painting is adding up the weight of a number of gold coins.

Notches

KEEPING ACCOUNTS
On tally sticks, the figures were cut into the stick in the form of a series of notches. The stick was then split in two along its length, through the notches, so each person involved in the deal had a record.

USING LOGARITHMS *below*
With logarithms and a slide rule it is possible to do complicated calculations very quickly.

Parallel scales

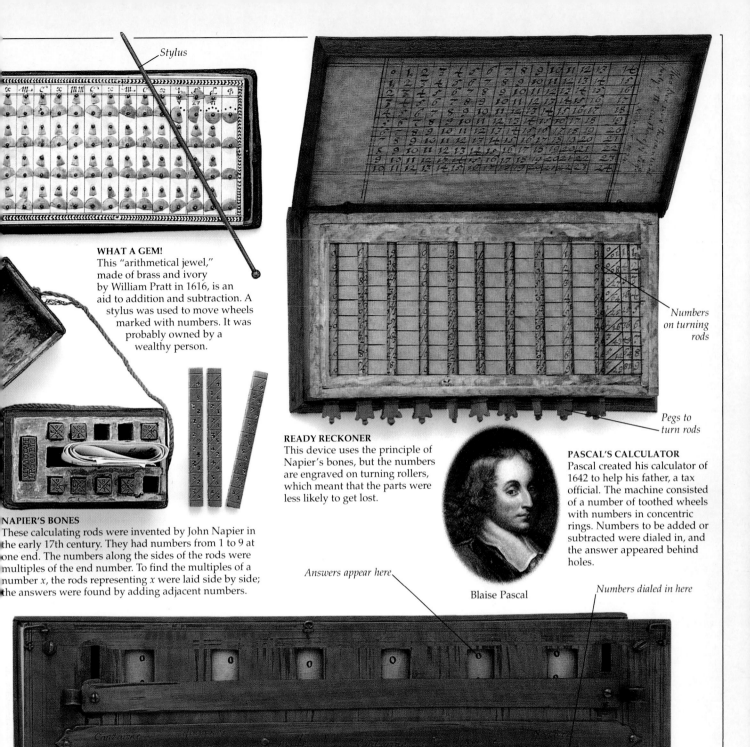

Stylus

WHAT A GEM!
This "arithmetical jewel," made of brass and ivory by William Pratt in 1616, is an aid to addition and subtraction. A stylus was used to move wheels marked with numbers. It was probably owned by a wealthy person.

Numbers on turning rods

Pegs to turn rods

READY RECKONER
This device uses the principle of Napier's bones, but the numbers are engraved on turning rollers, which meant that the parts were less likely to get lost.

NAPIER'S BONES
These calculating rods were invented by John Napier in the early 17th century. They had numbers from 1 to 9 at one end. The numbers along the sides of the rods were multiples of the end number. To find the multiples of a number x, the rods representing x were laid side by side; the answers were found by adding adjacent numbers.

Blaise Pascal

PASCAL'S CALCULATOR
Pascal created his calculator of 1642 to help his father, a tax official. The machine consisted of a number of toothed wheels with numbers in concentric rings. Numbers to be added or subtracted were dialed in, and the answer appeared behind holes.

Answers appear here

Numbers dialed in here

The steam engine

Hero of Alexandria's steam engine

THE POWER DEVELOPED BY STEAM has fascinated people for hundreds of years. During the first century A.D., Greek scientists realized that steam contained energy that could possibly be used by people. But the ancient Greeks did not use steam power to drive machinery. The first steam engines were designed at the end of the 17th century by engineers such as the Marquis of Worcester and Thomas Savery. Savery's engine was intended to be used for pumping water out of mines. The first really practical steam engine was designed by Thomas Newcomen, whose first engine appeared in 1712. Scottish instrument-maker James Watt improved the steam engine still further. His engines condensed steam outside the main cylinder, which conserved heat by dispensing with the need alternately to heat and cool the cylinder. The engines also used steam to force the piston down to increase efficiency. The new engines soon became a major source of power for factories and mines. Later developments included the more compact, high-pressure engine, which was used in locomotives and ships.

Parallel motion

Piston rod

Cylinder

GREEK STEAM POWER
Some time during the 1st century A.D., the Greek scientist Hero of Alexandria invented the æolipile – a simple steam engine that used the principle of jet propulsion. Water was boiled inside the sphere, and steam came out of bent jets attached it. This made the ball turn around. The device was not used for any practical purpose.

Valve chest

PUMPING WATER
English engineer Thomas Savery patented a machine for pumping water from mines in 1698. Steam from a boiler passed into a pair of vessels. The steam was then condensed back into water, creating a low pressure area and sucking water from the mine below. Using stop cocks and valves, steam pressure was then directed to push the water up a vertical outlet pipe. Thomas Newcomen, an English blacksmith, improved on this engine in 1712.

"Eduction pipe" to condenser

Air pump

Cistern containing condenser and air pump

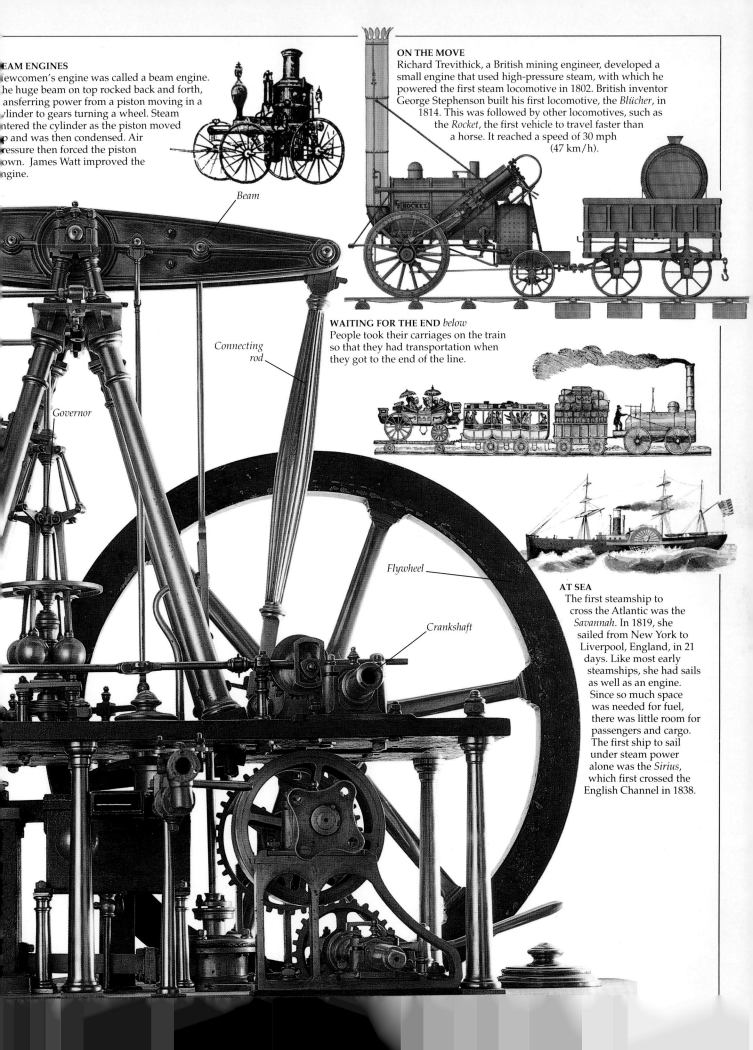

[B]EAM ENGINES
[N]ewcomen's engine was called a beam engine.
[T]he huge beam on top rocked back and forth,
[tr]ansferring power from a piston moving in a
[cy]linder to gears turning a wheel. Steam
[e]ntered the cylinder as the piston moved
[u]p and was then condensed. Air
[p]ressure then forced the piston
[d]own. James Watt improved the
[e]ngine.

Beam

ON THE MOVE
Richard Trevithick, a British mining engineer, developed a
small engine that used high-pressure steam, with which he
powered the first steam locomotive in 1802. British inventor
George Stephenson built his first locomotive, the *Blücher*, in
1814. This was followed by other locomotives, such as
the *Rocket*, the first vehicle to travel faster than
a horse. It reached a speed of 30 mph
(47 km/h).

*Connecting
rod*

Governor

WAITING FOR THE END *below*
People took their carriages on the train
so that they had transportation when
they got to the end of the line.

Flywheel

Crankshaft

AT SEA
The first steamship to
cross the Atlantic was the
Savannah. In 1819, she
sailed from New York to
Liverpool, England, in 21
days. Like most early
steamships, she had sails
as well as an engine.
Since so much space
was needed for fuel,
there was little room for
passengers and cargo.
The first ship to sail
under steam power
alone was the *Sirius*,
which first crossed the
English Channel in 1838.

Navigation and surveying

Chinese mariner's compass

18th-century English compass

THE MORE PEOPLE TRAVELED by boat, the more important the skills of navigation became. Navigation probably originated on the Nile and Euphrates rivers about 5,000 years ago when the Egyptians and Babylonians established trading routes. The Egyptians also pioneered surveying, essential for creating large buildings such as the pyramids. Navigation and surveying are related because both deal with measuring angles and calculating long distances. From around 500 B.C., first the Greeks, then the Arabs and Indians, established astronomy, geometry, and trigonometry as sciences and created such instruments as the astrolabe and compass. Understanding the movements of heavenly bodies and the relationship between angles and distances, medieval seafarers were able to create a system of longitude and latitude for finding their way at sea without reference to landmarks. The Romans pioneered the widespread use of accurate surveying instruments, and Renaissance architects added the theodolite, our most important surveying tool.

IN THE RIGHT DIRECTION
Magnetic compasses were used in Europe by about A.D.1200, but the Chinese are thought to have noticed about 1,000 years before that a suspended piece of lodestone (a magnetic iron mineral) points north-south.

Stones suspended from crossed sticks set at right angles to one another

Handle

RIGHT ANGLE *above*
Early surveyor's instruments such as the Egyptian groma were useful only on flat terrain and for setting a limited range of angles. With the groma, distant objects were marked out against the position of the stones in a horizontal plane.

Central arm

STRETCHING IT OUT
Ropes, chains, tapes, and rods have all been used for measuring distances. In about 1620, Edmund Gunter developed this type of metal chain for determining the area of plots of land. The chain is 66 ft (20 m) long and is made of 100 links. Markers are placed at regular intervals.

Brass marker

OCTANT
In the 1730s, English seafarer John Hadley invented the octant. This version is from about 1750. It enabled navigators to measure the altitude of the sun, moon, and stars so that they could find their latitude.

Chain link

SETTING BY THE SUN *above*
Medieval surveyors and navigators used instruments like the astrolabe (bottom right), the cross-staff (top right), and a measuring compass (left). The astrolabe was a 5th-century Arab development of ancient Greek astronomical instruments used to tell the local time by the position of the sun in the sky.

TURNING FULL CIRCLE
In 1676, Italian Joannes Macarius was so proud of this highly decorated circumferentor that he had his name engraved on it. It enabled the user to compare angles and figure out how far away a distant object was.

Three sets of degrees and angles on a graduated (divided) scale

Scales of length

Sight

Sight

Mirror

SMALL SEXTANT *above*
Sextants like this one from 1850 were used by army personnel and roadbuilders for making military maps and surveying land for roads or railways.

Telescopic sight

Ebony frame

Ivory scale

BURNING BRIGHT
The Pharos at Alexandria in Egypt was the first lighthouse and one of the seven wonders of the ancient world. Built in about 300 B.C., it stood 400 ft (122 m) tall. Its mirrors projected light from a giant fire to ships far out to sea.

STARRY-EYED
The octant was not good for working out longitude. In England in 1757, John Campbell developed the sextant for measuring both longitude and latitude.

Surveyor using a backstaff

Graduated angle scale

Reading marker

Sights

Scale measuring angles

HALFWAY HOUSE
The graphometer was a surveyor's instrument with a graduated half-circle. It was first described by Frenchman Phillipe Danfrie in 1597 and was a forerunner of the circumferentor.

Spinning and weaving

EARLY PEOPLE used animal skins to help them keep warm but about 10,000 years ago, people learned how to make cloth. Wool, cotton, flax, or hemp was first spun into a thin thread, using a spindle. The thread was then woven into a fabric. The earliest weaving machines probably consisted of little more than a pair of sticks that held a set of parallel threads, called the warp, while the cross-thread, called the weft, was inserted. Later machines called looms had rods that separated the threads to allow the weft to be inserted more easily. A piece of wood, called the shuttle, holding a spool of thread, was passed between the separated threads. The basic principles of spinning and weaving have stayed the same until the present day, though during the industrial revolution of the 18th century many ways were found of automating the processes. With new machines such as the spinning mule, many threads could be spun at the same time, and, with the help of devices like the flying shuttle, broad pieces of cloth could be woven at great speed.

CLOTHMAKING IN THE MIDDLE AGES
In about A.D. 1300, an improved loom was introduced to Europe from India. It was called the horizontal loom and had a framework of string or wire to separate the warp threads. The shuttle was passed across the loom by hand.

ANCIENT SPINDLE
Spindles like this were turned by hand to twist the fibers, and then allowed to hang so that the fibers were drawn in to a thread. This example was found in 1921 at the ancient Egyptian site at Tel el Amarna.

Drive thread

Wool

Wooden wheel

SPINNING AT HOME
The spinning wheel, which was introduced to Europe from India about A.D. 1200, speeded up the spinning process. In the 16th century, a foot treadle was added, freeing the spinner's hands – the left to draw out the fiber, the right to twist the thread.

SPINNING WHEEL
This type of spinning wheel, called the wool wheel, was used in homes until about 200 years ago. Spinning wheels like this, turned by hand, produced a fine yarn of even thickness.

FLYER right

About 250 years ago, many improvements were made to spinning machines. In 1769, Englishman Richard Arkwright introduced the flyer spinning frame. The flyer drew out the thread, then twisted it as it was wound onto a spool or bobbin. Also in England, ten years later, Samuel Crompton introduced the "spinning mule;" it could spin up to 1,000 threads at a time.

CHILD LABOR *above*

With the new machinery, weaving moved out of homes into factories, where water or steam power was available to work the machines. Young people were employed to crawl under machines to mend broken threads or clean up dust.

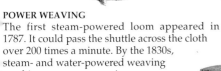

POWER WEAVING

The first steam-powered loom appeared in 1787. It could pass the shuttle across the cloth over 200 times a minute. By the 1830s, steam- and water-powered weaving machines were common in factories.

Fiber to be spun

Spun thread

Bobbins

Drive wheel

Batteries

OVER 2,000 YEARS AGO, the Greek scientist Thales produced small electric sparks by rubbing a cloth on amber, a hard yellow substance formed from the sap of long-dead trees. But it was a long time before people succeeded in harnessing this power in the form of a battery – a device for producing a steady flow of electricity. It was in 1800 that Alessandro Volta (1745-1827) published details of the first battery. Volta's battery produced electricity using the chemical reaction between certain solutions and metal electrodes. Other scientists, such as John Frederic Daniell (1790-1845), improved Volta's design by using different materials for the electrodes. Today's batteries follow the same basic design but use modern materials.

Metal electrodes

Fabric pads

VOLTA'S PILE *above*
Volta's battery, or "pile," consisted of disks of zinc and silver or copper separated by pads moistened with a weak acid or salt solution. Electricity flowed through a wire linking the top and bottom disks. An electrical unit, the volt, is named after Volta.

LIGHTNING FLASH
In 1752, inventor Benjamin Franklin flew a kite in a thunderstorm. Electricity flowed down the wet line and produced a small spark, showing that lightning bolts were huge electric sparks.

ANIMAL ELECTRICITY
Luigi Galvani (1737-1798) found that the legs of dead frogs twitched when they were touched with metal rods. He thought the legs contained "animal electricity." Volta suggested a different explanation. Animals do produce electricity, but the twitching of the frog's legs was probably caused by the metal rods and the moisture in the legs forming a simple electric cell.

Space filled with acid or solution

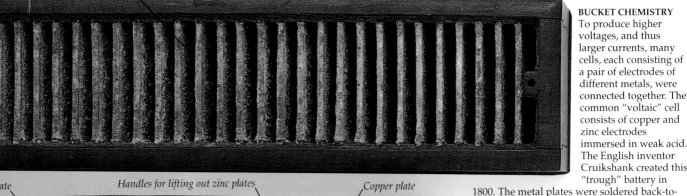

Zinc plate *Handles for lifting out zinc plates* *Copper plate*

1899·45

BUCKET CHEMISTRY
To produce higher voltages, and thus larger currents, many cells, each consisting of a pair of electrodes of different metals, were connected together. The common "voltaic" cell consists of copper and zinc electrodes immersed in weak acid. The English inventor Cruikshank created this "trough" battery in 1800. The metal plates were soldered back-to-back and cemented into slots in a wooden case. The case was then filled with a dilute acid or a solution of ammonium chloride, a salt.

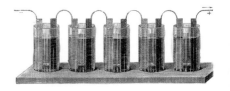

DIPPING IN, DRYING OUT
In about 1807, W. H. Wollaston, an English chemist, created a battery like this. Zinc plates were fixed between the arms of U-shaped copper plates, so that both sides of the zinc were used. The zinc plates were lifted out of the electrolyte to save zinc when the battery was not in use.

RELIABLE ELECTRICITY

The Daniell cell was the first reliable source of electricity. It produced a steady voltage over a considerable time. The cell had a copper electrode immersed in copper sulphate solution, and a zinc electrode in sulphuric acid. The liquids were kept separate by a porous diaphragm.

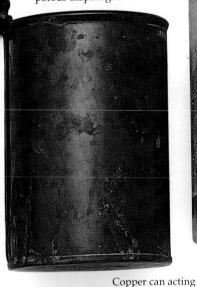

Porous diaphragm

Copper can acting as electrode

Terminal

RECHARGEABLE BATTERY

The French scientist Gaston Planté was a pioneer of the lead-acid accumulator, which can be recharged when it runs down. It has electrodes of lead and lead oxide in strong sulphuric acid.

Zinc rod electrode

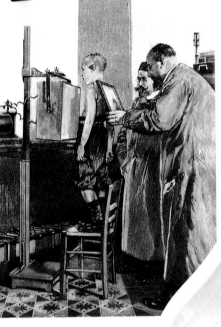

WILHELM ROENTGEN *right*
The German scientist Wilhelm Roentgen (1845-1923) discovered X-rays in 1895. Roentgen did not understand what these rays were so he named them X-rays.

GASSNER CELL *left*
Chemist Carl Gassner developed a pioneering type of "dry" cell. He used a zinc case as the negative (-) electrode, and a carbon rod as the positive (+) electrode. In between them was a paste of ammonium chloride solution and Plaster of Paris.

HUBBLE BUBBLE *right*
Some early batteries used concentrated nitric acid, but they gave off poisonous fumes. To avoid such hazards, the bichromate cell was developed in the 1850s. It used a glass flask filled with chromic acid. Zinc and carbon plates were used as electrodes.

HARVEY & PEAK,
Scientific Instrument Manufacturers
And Electridians,
6 SANDRINGHAM BUILDINGS,
CHARING CROSS ROAD,
LONDON, W.C

EVER READY
MADE IN BRITAIN

PATENT No. 536869

B103

EVER READY
UNIT CELL
U2

VIDOR
ETERNACELL
LITHIUM

VIDOR
ETERNACELL
LITHIUM

POWERPACKS *left*
The so-called "dry" cell has a moist paste electrolyte inside a zinc container which acts as one electrode. The other electrode is a carbon rod in the center of the cell. Small modern batteries use a variety of materials for the electrodes. Mercury batteries were the first long-life dry cells. Some batteries use lithium, the lightest of metals. They have a very long life and are therefore used in heart pacemakers.

Photography

THE INVENTION OF PHOTOGRAPHY made accurate images of any object rapidly available for the first time. It sprang from a combination of optics (see p. 28) and chemistry. The projection of the sun's image on a screen had been explored by Arab astronomers in the 9th century A.D., and by the Chinese before them. By the 16th century, Italian artists such as Canaletto were using lenses and a camera obscura to help them make accurate drawings. In 1725 a German anatomy professor, Johann Heinrich Schulze, noticed that a solution of silver nitrate in a flask turned black when stood in sunlight. In 1827 a metal plate was coated with a light-sensitive material, and a permanent visual record of an object was made.

IN THE BLACK BOX
The camera obscura (from the Latin for dark room) was at first just a darkened room or large box with a tiny opening at the front and a clear screen or wall at the back on which images were projected. From the 16th century, a lens was used instead of the "pinhole."

EGG ON ONE'S FACE
By 1841 Englishman William Henry Fox Talbot had developed the Calotype. This is an early example. It was an improved version of a process he had announced two years before, within days of Daguerre's announcement. It provided a negative image from which positives could be printed.

The daguerreotype

Joseph Nicéphore Niepce took the first surviving photograph. In 1827 he coated a pewter plate with bitumen and exposed it to light in a camera. Where light struck, the bitumen hardened. The unhardened areas were then dissolved away to leave a visible image. In 1839, his one-time partner, Louis Jacques Daguerre, developed a superior photographic process, producing the Daguerreotype.

EXPOSING THE PLATE *below*
In some daguerreotype cameras, the object was viewed through a hole in the back of the box. Then the photographic plate, protected by a cover, was slid into place. The lens cap and the cover were removed to expose the plate, then replaced.

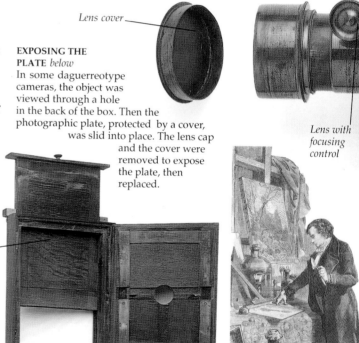

Lens cover

Lens with focusing control

Plate holder

DAGUERREOTYPE IMAGE
A daguerreotype consisted of a copper plate coated with silver and treated with iodine vapor to make it sensitive to light. It was exposed in the camera, then the image was developed by mercury vapor and fixed with a strong solution of ordinary salt.

Aperture rings

MAKING ADJUSTMENTS
By using screw-in lens fittings and different sized diaphragm rings to adjust the lens aperture, as on this folding daguerreotype camera of the 1840s, it became possible to photograph both close-up and distant objects in a variety of lighting conditions.

Lens and attachments

Folding daguerreotype camera

HEAVY LOADS
Enlargements could not be made with the early photographic processes, so for large pictures, big glass plates were used. With a dark tent for inspecting wet plates as they were exposed, plus water, chemicals, and plates, the equipment could weigh over 110 lb (50 kg).

The wet plate

From 1839 on, the pioneers of photography concentrated on the use of salts of silver as the light-sensitive material. In 1851, Frederick Scott Archer created a glass photographic plate more light-sensitive than its predecessors. It recorded negative images of fine detail with exposures of less than 30 seconds. The plate was coated with a chemical mix, put in the camera, and exposed while still wet. It was a messy process, but gave excellent results.

Chemicals for wet-plate process

Plate holder

Wet-plate negative

CHEMICALS *above right*
A wet plate consisted of a glass sheet coated with silver salts and a sticky material called collodion. It was usually developed with pyrogallic acid and fixed with sodium thiosulphate ("hypo"). Chemicals were dispensed from small bottles.

IN AND OUT OF VIEW
This wet-plate camera was mounted on a tripod. The rear section into which the photographic plate was inserted could slide toward or away from the front lens section to increase or decrease the image size and produce a clear picture. Fine focusing was by means of a knob on the lens tube.

Modern photography

In the 1870s dry gelatine-coated plates covered with extremely light-sensitive silver bromide were developed. Soon more sensitive paper allowed many prints to be made from a negative quickly and easily in a darkroom. In 1888 George Eastman introduced a small, lightweight camera. It used film which came on a roll.

PHOTOGRAPHY FOR ALL
In the early 1900s Eastman developed inexpensive Brownie box cameras such as this, and amateur photography was born. Each time a photo was taken, you would wind the film to be ready for the next shot.

Film winder

CANDID CAMERA *right*
By the 1920s German optical instrument manufacturers such as Carl Zeiss were developing small precision cameras. This 1937 single-lens reflex (SLR) Exakta model is in many ways the forerunner of a whole generation of modern cameras.

Viewfinder

Film winder

Lens

SLR camera

ROLL FILM
Eastman's early roll film consisted of a long thin strip of paper from which the negative coating was stripped and put down on glass plates before printing. In 1889 celluloid roll film came on the market. The light-sensitive emulsion was coated onto a see-through base so that the stripping process was eliminated.

Medical inventions

Steam generator

PEOPLE HAVE ALWAYS practiced some form of medicine. Early peoples used herbs to cure illnesses. Some prehistoric skulls have been found with round holes, probably drilled with a trepan, a surgeon's circular saw. The ancient Greeks used this operation to relieve pressure on the brain after severe head injuries. The ancient Chinese practiced acupuncture, inserting needles into one part of the body to relieve pain or the symptoms of disease in another part. But until well into the 19th century, a surgeon's instruments differed little from early ones – scalpels, forceps, various hooks, saws, and other tools to perform amputations or to extract teeth. The first instruments used to determine the cause of illnesses were developed in Renaissance Europe following the pioneering anatomical work of scientists such as Leonardo da Vinci and Andreas Vesalius. In the 19th century, medicine developed quickly; much of the equipment still used in medicine and dentistry today, from stethoscopes to dental drills, were developed at this time.

Carbolic acid reservoir

PLUNGING IN
Syringes were first used in ancient India, China, and North Africa. Nowadays, syringes consist of a hollow glass or plastic barrel and a plunger. A syringe fitted with a sharp hollow needle was first used in about 1850 by French surgeon Charles Gabriel Pravaz to inject drugs.

Flexible rubber tube

Porcelain teeth

Mouthpiece placed over patient's mouth had valves for breathing in and out

NUMBING PAIN
Before the discovery of anesthetics in 1846, surgery was done while the patient was still conscious and capable of feeling pain. Later, nitrous oxide (laughing gas), ether, or chloroform was used to numb pain. The gases were inhaled via a face mask.

Coiled spring

Ivory lower plate

YOU WON'T FEEL A THING
By the 1850s, anesthetics were used by dentists to "kill" pain. The first dental drills appeared in the 1860s.

DRILLING DOWN *right*
The Harrington "Erado" clockwork dental drill dates from about 1864. When fully wound, it worked for up to 2 minutes.

Drill bit

FIRM BITE *above*
The first full set of false teeth similar to those used today was made in France in the 1780s. This set of partial dentures dates from about 1860.

SPRAY IT ON *left*
By 1867 Scottish surgeon Joseph Lister had developed an antiseptic carbolic steam spray. It created a mist of carbolic acid intended to kill germs around the operation site. This version dates from about 1875.

THROUGH THE LOOKING TUBE *right*
Different types of endoscope, for viewing inside the body without surgery, were developed in the 19th century. This 1880s version used a candle as a light source.

Candle

DOWN THE TUBE
In 1819 French physician René Laennec created a tube through which he could hear the patient's heartbeat.

Ivory earpiece

Speculum - placed in patient's ear

Funnel for concentrating light

Viewing lens

LISTENING IN
Laennec's single-tube stethoscope was later developed into this 1855 version of the present-day design, with two earpieces. The stethoscope is used to listen to the sounds made by the heart, lungs, and blood vessels, and to the heartbeat of a baby in the womb.

TAKING THE PULSE *left*
In the early 17th century, physician William Harvey was the first to show how blood circulated around the body. But it was not until much later that the link between the pulse, heart activity, and health was established.

Ether vapor outlet valve

Air inlet valve

Metal tubes for transmitting the sounds. Today, tubes are made of plastic

UNDER PRESSURE *above*
Blood pressure is measured by feeling the pulse and slowly applying a measured force to the skin until the pulse disappears. The instrument that does this is called a sphygmomanometer and was invented by Samuel von Basch in 1891.

HOT UNDER THE COLLAR? *right*
These thermometers, from about 1865, were placed in the mouth (straight version) or under the armpit (curved-end type). Measuring the patient's temperature was not common practice until the early decades of this century.

Temperature scale in degrees Fahrenheit

Cone

Reservoir of mercury

Ether-soaked sponges

LIGHT-HEADED FEELING
In the 19th century, ether was used as an anesthetic. The "Letheon" ether inhaler of 1847 comprised a glass jar filled with ether-soaked sponges through which air was drawn as the patient breathed in.

Kink in tube - to give good fit in armpit

HOLLOW SOUNDS *right*
The disk-shaped sound collector on this 1830s wooden stethoscope would have been used to listen to high-pitched sounds, such as those made by the lungs, rather than low-pitched ones, such as heartbeats.

The telephone

FOR CENTURIES, people have tried to send signals over long distances using bonfires and flashing mirrors to carry messages. It was the Frenchman Claude Chappe who in 1793 devised the word "telegraph" (literally, writing at a distance) to describe his message machine. Moving arms mounted on towertops signaled numbers and letters. Over the next 40 years, electric telegraphs were developed. And in 1876 Alexander Graham Bell invented the telephone, enabling speech to be sent along wires for the first time. Bell's work with the deaf led to an interest in how sounds are produced by vibrations in the air. His research on a device called the "harmonic telegraph" led him to discover that an electric current could be changed to resemble the vibrations made by a speaking voice. This was the principle on which he based the telephone.

OPENING SPEECH
Alexander Graham Bell (1847–1922) developed the telephone after working as a speech teacher with deaf people. Here he is making the first call on the New York to Chicago line.

MAKING A CONNECTION
These two men are using early Edison equipment to make their telephone calls. Each has a different arrangement - one is a modern-style receiver and the other, a two-piece apparatus for speaking and listening. All calls had to be made via the operator.

Magnet

Earpiece and mouthpiece combined

ALL-IN-ONE
Early models like Bell's "Box telephone" of 1876-1877 had a trumpet-like mouthpiece and earpiece combined. The instrument contained a membrane that vibrated when someone spoke into the mouthpiece. The changing vibration varied an electric current in a wire, and the receiver turned the varying current back into vibrations that the listener could hear.

Wire coil

Iron diaphragm

EARPIECE
In this earpiece of about 1878, a fluctuating electric current passing through the wire coil made the iron diaphragm move to make sounds.

The telegraph

The telegraph, the forerunner of the telephone, allowed signals to be sent along a wire. The first telegraphs were used on the railroads to help keep track of trains. Later, telegraph wire linked major cities.

MESSAGE MACHINES
With the Morse key (left) you could send signals made up of short dots and long dashes. In the Cooke and Wheatstone system (right), the electric current made needles point at different letters.

DON'T HANG UP
In 1877 Thomas Edison developed different mouthpiece and earpiece units. Models such as this were hung from a special switch that disconnected the line on closing.

WIRED FOR SOUND
The first telephone cables used copper wires sheathed in glass. Then iron wires were used for their strength – especially for overhead lines.

EASY LISTENING
This wall-mounted telephone of 1879 was invented by Thomas Edison and has a microphone and receiver of his design. The user had to wind the handle while listening. A ring of the bell indicated an incoming call or a successful connection.

Earpiece

REPEAT THAT NUMBER
The earliest telephone exchanges were manual. One of the dozens of operators took your number and the number you wanted, and plugged in your line wire to complete the appropriate electrical circuit.

HANDSETS
By 1885 the transmitter and receiver had been combined to form a handset. At first this was metal, but by 1929 plastic handsets were common.

Mouthpiece

Mouthpiece

Hook for earpiece

Transmitter containing carbon granules, compressed and released by sound waves to create an electric current of varying strength

IT'S A STICK-UP
Some candlestick-shaped phones of the 1920s and 1930s had a dial for calling numbers via an automatic exchange.

Numbered dial

Earpiece

LONG DISTANCE CALL FOR YOU
"Cradle" telephones like this were popular by the 1890s. This one dates from 1937, by which time there was a transatlantic telephone service between London and New York.

Drawer for directory

Recording

SOUNDS WERE RECORDED for the first time in 1877 on a telephone repeater devised by Thomas Edison. This device recorded sound vibrations as indentations in a sheet of paper passing over a rotating cylinder. Edison first tried out his machine by shouting "Hello" into the mouthpiece. When the paper was pulled beneath a stylus attached to a diaphragm, the word was reproduced. This mechanical-acoustic method of recording continued until electrical systems appeared in the 1920s. Tape recording systems were developed using the principles of magnetics. These systems received a commercial boost, first in 1935, with the development of magnetic plastic tape, and then in the 1960s, with the use of microelectronics (p. 62).

TWO IN ONE MACHINE
By 1877, Edison had created separate devices for recording and playing back. Sounds made into a horn caused its diaphragm to vibrate and its stylus to create indentations on a thin sheet of tinfoil wrapped around the recording drum. Putting the playback stylus and its diaphragm in contact with the foil and rotating the drum reproduced the sounds via a second diaphragm.

Mouthpiece (horn not shown)

Drive axle, threaded to move length of foil beneath fixed stylus

Tinfoil was wrapped around this brass drum

Cross-section showing needle on cylinder

Edison phonograph showing positions of needle and horn

Position of horn

PLAY IT AGAIN, SAM
The playback mechanism comprised a stylus made of steel in contact with a thin iron diaphragm. The wooden mount was flipped over so the stylus made close contact with the foil as it rotated. Vibrations from the foil were transferred to the diaphragm. As the diaphragm moved in and out, it created sound waves.

Cylinder
and box

78 rpm
record

IN THE GROOVE *above*
Chichester Bell's sapphire needles cut a continuous groove in a wax cylinder, the depth of which varied with the intensity of the sound being recorded. These later cylinder recordings lasted for up to 4 minutes.

Needles

WAXING LYRICAL *left*
Edison's tinfoil recordings played for only about a minute and were soon worn out by the steel needles. In the mid 1880s, Chichester Bell, cousin of the inventor of the telephone, with scientist Charles Tainter, used a sapphire stylus and developed a wax-coated cylinder as a more durable alternative. Edison created this version in about 1905.

ON THE FLAT
In 1888 Emile Berliner created the forerunner of modern records and record players. The playback mechanisms were similar to their predecessors, but instead of a cylinder, Berliner used a flat disk with a groove that varied not in depth but in side-to-side movement.

CUTTING A RECORD *above*
Berliner's first record system used a glass disk coated with hardened shellac as a "negative." This was used to photoengrave the recording pattern onto flat-metal-disk "positives." In 1895 he developed a method used until recently – shellac positives, like this 78 rpm record, were pressed from a nickel-plated negative.

Horn to channel sounds from the iron diaphragm

Steel needle

Turntable

Tape recording

In 1898 Danish inventor Valdemar Poulsen produced the first magnetic recorder. Recordings were made on steel piano wire. In the 1930s two German companies, Telefunken and I. G. Farben, developed a plastic tape coated with magnetic iron oxide, which soon replaced steel wires and tapes.

WIRED UP *left*
This 1903 Poulsen telegraphone was electrically driven and replayed. The machine was used primarily for dictation and telephone message work. The sounds were recorded on wire.

ON TAPE *above*
This tape recorder of about 1950 has three heads, one to erase previous recordings, one to record, and the third to replay.

The internal combustion engine

THE INTERNAL COMBUSTION ENGINE created a revolution in transportation almost as great as that caused by the wheel. For the first time, a small, relatively efficient engine was available, leading to the production of vehicles from cars to aircraft. Inside an internal combustion engine a fuel burns (combusts) to produce power. The fuel burns inside a tube called a cylinder. When the fuel burns, hot gases form and push a piston down the cylinder. The piston's movement produces the power to drive wheels or machinery. The first working internal combustion engine was built in 1860 by Belgian inventor Etienne Lenoir. It was powered by natural gas. The German engineer Nikolaus Otto built an improved engine in 1876. This engine's power was produced by four movements of the piston, and it became known as the four-stroke engine. The four-stroke engine was developed by Gottlieb Daimler and Karl Benz, leading to the production of the first automobile in 1885.

FIRST CAR
Daimler and Benz adapted Otto's engine so that it could run on gasoline, a more useful fuel than natural gas. This meant that the engine was not tied to the gas supply and had enough power to drive a passenger-carrying vehicle.

Exhaust manifold

Cooling fan

MODIFIED STEAM ENGINE *left*
This engine from the 1890s was halfway between a steam engine and a modern gasoline engine. It had a slide valve system alongside the cylinder, like a steam engine. The slide valve allowed the burned fuel to escape as the piston pushed it from the cylinder.

NON-STARTER
In this unsuccessful 1838 design for an internal combustion engine the fuel was burned inside a cylinder, which rotated as hot gas escaped through vents.

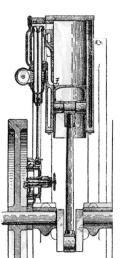

GAS ENGINE
In Lenoir's engine of 1860, a mixture of coal gas and air was drawn into the cylinder by the movement of the piston. The mixture was then ignited by an electrical spark, and the exploding gas forced the piston to the end of the cylinder.

Camshaft

Crankshaft

FOUR-STROKE CYCLE

During the "induction" stroke, the piston moves down, sucking the fuel-air mixture into the cylinder through the open inlet valve. During the "compression" stroke, the piston moves up, compressing the mixture; the spark plug ignites the mixture at the top of the stroke. During the "power" stroke, the expanding gases (the burned fuel) push the piston down. During the "exhaust" stroke, the piston moves up, forcing the hot gasses out through the open exhaust valve.

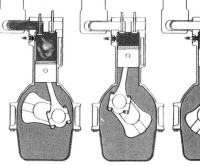

Induction   Compression   Power   Exhaust

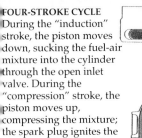

CAR OF THE PEOPLE *right*

The 1908 Model T Ford was the first car to be mass-produced. Over 15 million were made before production ended in 1927. By 1910, the main features of many later cars had been established: a four-stroke engine mounted at the front with power being transmitted to the rear wheels via a drive shaft.

Valve

Cylinder

Piston

Gudgeon

Connecting rod

Clutch

INSIDE AN ENGINE

This 1925 Morris engine is a basic power unit for a family car. Its four in-line cylinders have aluminum pistons. The valves are opened by push rods operated by a camshaft and closed by springs. Power is transmitted via the crankshaft to the gearbox. The clutch disconnects the engine from the gearbox when the driver changes gear.

Cinema

I N 1824, AN ENGLISH DOCTOR, P. M. Roget, first explained the phenomenon of "persistence of vision." He noticed that if you see an object in a series of closely similar positions in a rapid sequence, your eyes tend to see a single moving object. It did not take people long to realize that a moving image could be created with a series of still images, and within 10 years scientists all over the world were developing a variety of devices for creating this illusion. Most of these machines remained little more than novelties or toys, but combined with improvements in illumination systems for magic lanterns and with developments in photography, they helped the progress of cinema technology. The first successful public showing of moving images created by cinematography was in the 1890s by two French brothers, Auguste and Louis Lumière. They created a combined camera and projector, the Cinématographe, which recorded the pictures on a celluloid strip.

ROUND AND ROUND
In the late 1870s, Eadweard Muybridge designed the zoopraxiscope for projecting moving images on a screen. The images were a sequence of pictures based on photographs, painted on a glass disk, which rotated to create a moving picture.

Slide holder

Lens

MAGIC LIGHT SHOW *above*
In a magic lantern, images on a transparent slide are projected onto a screen using a lens and a light source. Early magic lanterns used a candle; later, limelight or carbon arc lamps were used to give more intense illumination.

MOVING PICTURES
The Lumières were among the first to demonstrate projected moving pictures. Their Cinématographe worked like a magic lantern but projected images from a continuous strip of film.

SILVER SCREEN
The Lumières' system was used for the first regular film shows in Europe. The brothers opened a theater in a café basement in 1895.

Light hood to prevent stray light reaching lens

An early movie-maker at work

Light-proof wooden film magazine

STRIP FEATURES *above*

In the 1880s, Muybridge produced thousands of sequences of photographs that showed animals and people in motion. He placed 12 or more cameras side by side and used electromagnetic shutters that fired at precise split-second intervals as the subject moved in front of them.

LONG AND WINDING PATH *right*

Movie film must be wound through the camera and projector at between 16 and 24 frames a second. Many yards of film are needed for shows lasting more than a few minutes. This English camera from 1909 had two 400-ft (120-m) film magazines. Film comes out of the first magazine, passes through the gate, and is fed into the lower magazine.

Film spools

GLORIOUS TECHNICOLOR

Color movies became common in the late 1940s. This Technicolor three-strip camera of 1932 has a prism beamsplitter behind the lens which exposes three separate negative films – each one sensitive to red, blue, or green light. The three images were dyed and then combined to make a single full-color print film for projecting.

Prism beamsplitter

Viewfinder

Film gate

Film-housing door opened to reveal film-winding system and film gate

Film revolution counter

Radio

GUGLIELMO MARCONI, experimenting in his parents' attic near Bologna, Italy, developed the first radio. Fascinated by the idea of using radio waves to send messages through the air, he created an invention that was to change the world, making wireless communication over long distances possible and transforming the entertainment business. For a transmitter he used an electric spark generator invented by Heinrich Hertz. Radio waves from this transmitter were detected by a "coherer," the invention of Frenchman Edouard Branly. The coherer turned the radio waves into an electric current. By sending radio signals across the room, Marconi made an electric bell ring. That was in 1894. Within eight years he was sending radio messages 3,000 miles (4,800 km) across the Atlantic.

A BRIGHT SPARK *above*
In 1888 Heinrich Hertz, a German physicist, made an electric spark jump between pairs of metal spheres, creating a current in a circuit nearby. Hertz was studying electromagnetic waves, a type of radiation that includes visible light, radio waves X-rays, infrared waves, and ultraviolet light.

Glass bulb

Positive electrode (anode)

Grid

Filament (negative electrode - cathode)

HEATING UP
Early radio receivers were not very sensitive. In 1904 Englishman John Ambrose Fleming first used a diode (a device with two electrodes) as a better detector of radio waves. It was a type of electron tube (p. 56). Diodes convert alternating electric currents into direct ones, for use in electric circuits.

CARRIER WAVES
Electron tubes like this triode of 1908 had a third electrode, the grid, between the filament and the positive electrode. These tubes allow telephone messages and microphone signals to be amplified. The amplified signals are combined with special radio waves known as carrier waves so that they can be transmitted over great distances .

IT'S THE CAT'S WHISKERS
When radio stations first started broadcasting in the early 1920s, listeners tuned-in using receivers made up of silicon crystals or lead compounds and thin wires popularly known as cat's whiskers. The radio signals were weak, so headphones were used. Headphones contain a pair of loudspeakers, which convert varying electric currents – the radio signals – into sound waves.

Electrical connections to battery

ACROSS THE AIRWAVES
Marconi developed radio as the first practical system of wireless telegraphy, which made possible uninterrupted communication over land and sea.

HEAVY SOUNDS
Electron tubes and other radio components needed direct current. Early radio sets ran off large powerful batteries. The resulting radio receiver was big and heavy. A separate loudspeaker was used with this model.

Coils

Tuning condenser

Tubes

Tuning dials

Volume control

Cat's whisker

Crystal

WHAT THE WHISKER DID
On this type of radio receiver, the crystal detector only worked when the cat's whisker made a point contact with the crystal. It was often difficult to establish contact, so crystal sets were not easy to use. They were soon superseded by sets using electron tubes.

WORDS AND PICTURES
In the 1920s, tubes like this triode not only enabled the first speech to be broadcast from England to Australia – by Marconi in 1924 – but also aided the development of television cameras, transmitters, and receivers.

Plug-in base

GOOD RECEPTION
This early tube receiver had a loudspeaker built into the cabinet.

RADIO COMES TO EVERY HOME
By the 1920s, many radio transmitters had been built, and radio was within reach of many households in the U.S. and Europe.

GATHER ROUND
This detail from a painting by W. R. Scott shows people gathering around a radio receiver at a Christmas party. In 1922, when this picture was painted, radio was still a new attraction for most people.

Inventions in the home

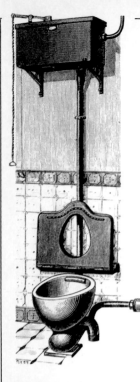

SCIENTIST MICHAEL FARADAY discovered how to generate electricity in 1831. But it was many years before electricity was used around the home. At first, large houses and factories installed their own generators and used electricity for lighting. The electric light bulb was first demonstrated in 1879. In 1882, the first large electric power station was built in New York. Gradually, as people began to realize how appliances could cut down on work in the home, mechanical items, such as early vacuum cleaners, were replaced by more efficient electrical versions. As the middle classes came to rely less and less on domestic servants, labor-saving appliances became more popular. Electric motors were applied to food mixers and hair dryers around 1920. Electric kettles, ovens, and heaters, making use of the heating ability of an electric current, had also appeared by this time. Some of these items were very similar in design to those used today.

HANDY FLUSH
The first description of a flush toilet was published by Sir John Harrington in 1596. But the idea did not catch on widely until drainage systems were installed in major cities. London's drainage system, for example, was not in operation until the 1860s. By this time several improved versions of the toilet had been patented.

KEEPING COOL
Electric refrigerators began to appear in the 1920s. They revolutionized food storage.

TEA'S UP
In the automatic tea maker of 1904, levers, springs, and the steam from the kettle activate stages in the teamaking process. A bell rings to tell you that the tea is ready.

ON THE BOIL
The Swan electric kettle of 1921 was the first with a totally immersed heating element. Earlier models had elements in a separate compartment in the bottom of the kettle, which wasted a lot of heat.

THE "WILSON" COOKER is Perfection for Baking Bread Pastry and Tea Cakes

COOK'S FRIEND
Before the 19th century, you had to light a fire to cook food. By 1879, an electric cooker had been designed in which food was heated by electricity passing through insulated wire wound around the cooking pot. In the 1890s, heating elements were made as iron plates with wires beneath. The modern element, which can be bent into any shape, came into use in the 1920s.

Heating element

EASY MIXING
The 1918 food mixer had two blades driven by an electric motor. A hinge allowed the mixer to be turned to a horizontal position.

Dowsing bulb

Electric motor

GOOD GROOMING
The 1925 electric hair dryer had a simple heater and a small fan. It was made of aluminum and had a wooden handle. The user had a choice of two heat settings.

KEEPING WARM
Early electric heaters used the Dowsing bulb. This was like an oversized light bulb, which was coated on the outside and mounted in front of a reflector in an attempt to concentrate the heat given off.

Heating element

ELECTRIC IRON
The first electric iron was heated by an electric arc between carbon rods and was highly dangerous. A safer iron was patented in 1882. It used an electrically heated wire element like the coils on a stove.

THE SAD IRON *left*
The most common form of iron in use from the 18th century until the early 20th century was the sad iron ("sad" meant heavy). These were used in pairs, with one heating up over the embers of a fire while the other was being used.

Bellows

QUICK COOKING *left*
The pressure cooker was invented by Frenchman Denis Papin in 1681. He called it the "new digester." Super-heated steam at high pressure formed inside the strong container. The high temperature cooked the food in a very short time.

CLEANING UP *right*
The mechanical vacuum cleaner of the early 20th century needed two people to operate it. A bellows was worked by a wooden handle, sucking dirt in. William Hoover began to make electric cleaners in 1908.

The cathode ray tube

In 1887, PHYSICIST William Crookes was investigating the properties of electricity. He used a glass electron tube containing two metal plates, the electrodes. When a high voltage was applied and the air pumped out of the tube, electricity passed between the electrodes and caused a glow in the tube. As the pressure fell (approaching a vacuum) the light went out, yet the glass itself glowed. Crookes called the rays which caused this cathode rays; they were, in fact, an invisible flow of electrons. Later, Ferdinand Braun created a tube with an end wall coated with a substance that glowed when struck by cathode rays. This was the forerunner of the modern TV receiver tube.

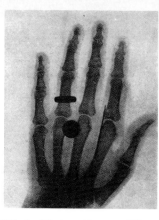

Wilhelm Röntgen discovered X-rays using a similar tube to Crookes in 1895.

Cathode – emitted electrons

Metal plates – one attracted the beam, the other repelled it

Anode with hole to create beam of electrons

Screen – coated with powder that glowed when struck by the beam

Electron gun

DOWN THE TUBE
Braun's 1897 tube incorporated two pairs of flat metal plates arranged at right angles to each other. The screen was coated with phosphorescent powder. By applying a voltage to the plates, Braun directed the beam of electrons (named cathode rays, because they were given off from the cathode) to create a bright spot of light on the screen. By varying the voltage across the plates, he could make the spot move around.

COLOR COMBINATION *below*
In 1953 a color television system was developed using a cathode ray tube with three electron guns – one each for blue, red, and green light – and a shadow mask, a stencil that directs each beam onto corresponding phosphor dots on the screen.

Induction coil to produce high voltage

Photographic plate recording X-rays passing through a hand

UNKNOWN FACTOR *left*
German physicist Wilhelm Röntgen noticed that as well as cathode rays, another form of radiation was emitted from a discharge tube when very high voltages were used. Unlike cathode rays, these rays, which he called X, for unknown, were not deflected by electrically charged plates or by magnets. They passed through materials and darkened photographic plates.

IN A SPIN *right*
In 1884 Paul Nipkow invented a system of spinning disks containing holes arranged in spiral form, which transformed an object into an image on a screen. In 1926 Scottish inventor John Logie Baird (seated in the picture) used Nipkow disks to give the world's first demonstration of television.

Single-beam gun

Electromagnetic coil to direct electron beams

CHEAPER TV
In the late 1960s, the Japanese firm Sony developed and patented the Trinitron system, a cathode ray tube with a different design from RCA's original color tube. This meant that they did not have to pay fees to RCA for every tube they made.

Electron gun producing 3 separate beams

Trinitron tube

TELEVISION GOES PUBLIC
In 1936 the BBC started the first public high-definition television service from this studio at Alexandra Palace, London. At first they used both Baird's system and one using the cathode ray tube. The latter gave the best results and Baird's system was never used again. In 1939 RCA started America's first fully electronic television service.

FASTER THAN THE EYE *below*
Until the 1960s, most home television receivers produced black and white pictures and operated with tubes (p. 52). The tube consisted of a single electron gun producing a beam that was made to scan the screen more than 50 times a second. As techniques improved, the length of the tube was shortened.

Phosphor screen

IN FRONT OF THE BOX *above*
Early television sets, such as this RCA Victor model, had small screens but contained such a mass of additional components that they were housed in large boxes. At the time, many such sets cost as much as a small car.

Electron gun

Flight

THE FIRST CREATURES to fly in a humanmade craft were a cockerel, a duck, and a sheep. They were sent up in a hot-air balloon made by the French Montgolfier brothers in September 1783. When the animals landed safely, the brothers were encouraged to send two of their friends, Pilâtre de Rozier and the Marquis d'Arlandes, on a 25-minute flight over Paris. Among the earliest pioneers of powered flight were Englishmen William Henson and John Stringfellow, who built a model aircraft powered by a steam engine in the 1840s. We do not know whether it flew or not – it may well have failed because of the heavy weight and low power of the engine. But it did have many of the features of the successful airplane. It was the Wright brothers who first achieved powered, controlled flight in a full-size aeroplane. Their *Flyer* of 1903 was powered by a lightweight gasoline engine.

AIRBORNE CARRIAGE
Henson and Stringfellow's "Aerial steam carriage" had many features that were used by later aircraft designers. It had a separate tail with rudders and elevators, and upward-sloping wings. The craft looks strange, but it was a surprisingly practical design.

Wooden and canvas wing

FIRST FLIGHT *below*
On June 4, 1783, Joseph and Etienne Montgolfier demonstrated a paper hot-air balloon. It climbed to about 6,000 ft (1,860 m). Later in the same year, the brothers sent up their animal and human passengers.

MECHANICAL WING
Some 500 years ago, Leonardo da Vinci designed a number of flying machines, most of which had mechanical flapping wings. They were bound to fail because of the great effort needed to flap the wings. Leonardo also designed a simple helicopter.

GLIDING FREE *above*
The first piloted glider was built by German engineer Otto Lilienthal. He made many flights between 1891 and 1896, when he was killed as his glider crashed. His work showed the basics of controlling a craft in the air.

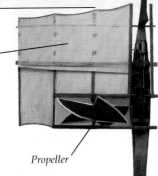

Wing

Propeller

MASTER OF THE WORLD? *right*
This design for a flying machine appeared in the book *Master of the World* by Jules Verne. Verne was vague about the power source and his design was generally impractical.

GETTING UP STEAM *left*
The model airplane made by Henson and Stringfellow had a specially made lightweight steam engine – the only type available at the time – to drive the twin propellers.

Housing for steam engine

IN CONTROL *above*
The brothers Wilbur and Orville Wright spent three years experimenting with gliders, learning how to control the craft. On the *Flyer*, the pilot lay on the lower wing and twisted the wings to roll the craft right or left. The craft also had elevators (for climbing and diving) and rudders (to control right and left turns).

FIRST POWERED FLIGHT
On December 17, 1903, the *Flyer* took off near Kitty Hawk, North Carolina, with Orville Wright as pilot. The machine rose to a height of 10 ft (3 m), and

landed heavily after 12 seconds. The brothers made three other flights that day. The longest lasted 59 seconds and covered 850 ft (260 m).

Plastics

PLASTICS ARE MATERIALS which can easily be formed into different shapes. They were first used to make imitations of other materials, but it soon became clear that they had qualities that were useful in other ways. Plastics are made up of long, chain-like molecules formed by a process (called polymerization) that joins small molecules together. The resulting long molecules give plastics their special properties. The first plastic, Parkesine, was made by modifying cellulose, a chainlike molecule, found in most plants. The first truly synthetic plastic was Bakelite, which was invented in 1909. The chemists of the 1920s and 1930s developed ways of making plastics from substances found in oil. Their efforts resulted in a range of materials with different heat, electrical, optical, and moulding properties. Plastics such as polyethylene, nylon, and acrylics, are widely used today.

IMITATION IVORY
Early plastics often had the appearance and feel of ivory, and carried names such as Ivoride. Materials like this were used for knife handles and combs.

IN FLAMES
In the 1860s, a plastic called Celluloid was developed. It was used as a substitute for ivory to make billiard balls, and for small items like this powder box. The new material made little impact at first, but in 1889 George Eastman began using it as a base for photographic film. Unfortunately, it had the disadvantage that it easily caught fire and sometimes exploded.

AROUND THE HOUSE
Plastics of the 1920s and 30s, like urea formaldehyde, were tough and nontoxic, and could be made any color with synthetic pigments. They were used for boxes, clock cases, piano keys, and lamps.

THE FIRST PLASTIC *right*
In 1862 Alexander Parkes made a hard material that could be molded into shapes. Called "Parkesine," it was the first semi-synthetic plastic.

Celluloid box

Hard, smooth surface

HEAT-PROOF
Leo Baekeland, a Belgium-born chemist working in the U.S., made a plastic from chemicals found in coal tar. His plastic, which he called Bakelite, was different from earlier plastics because it could only be softened by heat once before it set hard.

Heat-proof Bakelite container

Marble-effect surface

Polyethylene eyeglasses

Film

Styrofoam
egg box

Imitation sponge

Nylon thread

PLASTIC FOAM *above*
Polystyrene was first made in the 1920s. It comes in two forms: a hard form and a lightweight foam full of small holes called Styrofoam.

NYLON ROPE
Nylon provides great strength in a narrow thickness, making it ideal for rope.

Molded
polyethylene
spade and
toy racket

Buttons
and pen

*Separate
nylon fibres*

Toy bricks

SHAPES AND SIZES
Plastic can be formed into intricate shapes, like this fine netting.

PLASTIC FIBERS *left*
It was chemist Wallace Carothers who produced a plastic called nylon in 1934. It was like artificial silk and could be drawn out into thin threads and woven into cloth or twisted around to create rope as strong as steel cable. Other artificial fibers, such as polyesters, were discovered in 1941. Polyester fibers are woven into cloth for shirts, trousers, and dresses.

Plastic wrench

Polyethylene
flower

The silicon chip

Early radios and television sets used electron tubes (p. 56) to manipulate their electric currents. These tubes were large, had a short life, and were costly to produce. In 1947, scientists at the Bell Telephone Laboratories invented the smaller, cheaper, and more reliable transistor to do the same job. With the development of spacecraft, still smaller components were needed, and by the end of the 1960s thousands of transistors and other electronic components were being crammed onto chips of silicon only 0.2 in (5 mm) square. These chips were soon being used to replace the mechanical control devices in products ranging from dishwashers to cameras. They were also taking the place of the bulky electronic circuits in computers. A computer that once took up a whole room could now be contained in a case that would fit on top of a desk. A revolution in information technology followed, with computers being used for everything from playing games to administering government departments.

BABBAGE'S ENGINE
The ancestor of the computer was Charles Babbage's "Difference Engine," a mechanical calculating device. Today tiny chips do the job of such cumbersome machines.

Silicon wafer containing several hundred tiny chips

Matrix of connections to be produced

‡ 4161 RC   ALUMINIUM

Silicon chip

Plastic housing

SILICON CRYSTAL
Silicon is usually found combined with oxygen as silica, one form of which is quartz. Pure silicon is dark gray, hard, and nonmetallic, and it forms crystals.

MAKING A CHIP
The electrical components and connections are built up in layers on a wafer of pure silicon 0.02 in (0.5 mm) thick. First, chemical impurities are embedded in specific regions of the silicon to alter their electrical properties. Then, aluminum connections (the equivalent of conventional wires) are laid on top.

CHIP OFF THE OLD BLOCK
In the early 1970s, different types of chip were developed to do specific jobs – such as memory chips and central processing chips. Each silicon chip, a few millimeters square, is mounted in a frame of connections and pins, made of copper coated with gold or tin. Fine gold wires link connector pads around the edge of the chip to the frame. The whole assembly is housed in a protective insulating plastic block.

Printed circuit board connectors

WIRED TOGETHER
On a printed circuit (PC) board, a network of copper tracks is created on an insulating board. Components, including silicon chips, are plugged or soldered into holes in the PC board.

OUT IN SPACE *below*
Computers are essential for spacecraft like this satellite. The silicon chip means that control devices can be housed in the limited space on board.

DESK-TOP BRAIN
The late 1970s saw the computer boom. Commodore introduced the PET, one of the first mass-produced personal computers. It was used mainly in businesses and schools.

Visual display terminal (VDT)

Keyboard

ON THE RIGHT TRACK
Under a microscope, the circuitry of a chip looks like a network of aluminum tracks and islands of silicon, treated to conduct electricity.

MAKING CONNECTIONS
A close-up shows the connector wires attached to the silicon. Robots have to be used to join the wires to the chip since the components are so tiny and must be very accurately positioned.

SMART CARD
These plastic cards contain a silicon chip programmed with details of your bank account. Each time a transaction is made, the card's processor carries out its own security and credit limit check and instantly balances the account.

Silicon chip

TELECARTE
50 UNITES

Index

A

abacus, 30
acupuncture, 42
adze, 10, 11
airplanes, 58, 59
amplification, 52
anesthetics, 42, 43
antiseptics, 43
Archer, Frederick Scott, 41
Archimedes, 8, 9, 24
Arkwright, Richard, 37
astronomy, 34
automobile, 12, 48

B

Babbage, Charles, 62
Baekeland, Leo, 60
balances, 16, 23
ballpoint pen, 18, 19
batteries, 38-39, 52, 53
beam engine, 33
bearings, 12, 13
Bell, Alexander Graham, 44
Bell, Chichester, 47
Benz, Karl, 48
Berliner, Emile, 47
bicycle, 12
binoculars, 29
Biro, Josef and Georg, 19
blast furnace, 15, 25
Braun, Ferdinand, 56
bronze, 10, 11, 14, 15, 16

C

calculators, 30, 31, 62
calipers, 17
Calotype, 40
camera, 40, 41, 51, 62
Campbell, John, 35
can opener, 6
Canaletto, 27
candles, 20, 21, 22
carbon arc, 50
carbon-filament lamp, 7
cast iron, 15
cathode rays, 56, 57
cat's whisker radio, 52
Celluloid, 41, 50, 60
cinematography, 50-51
circumferentor, 35 clocks,

12, 22-23
cloth, 36, 37
compass, 34, 35
compositor, 27
computers, 62, 63
copper, 14, 38, 39, 44, 63
cranes, 24
Crompton, Samuel, 37
Crookes, William, 56

D

da Vinci, Leonardo, 42, 58
Daguerre, Louis Jacques, 40
Daimler, Gottlieb, 48
Danfrie, Phillipe, 35
Daniell, John Frederic, 38
de Rozier, Pilâtre, 58
diode, 52
drills, 8-9, 42
Dystrop wheels, 13

E

Eastman, George, 41, 60
Edison, Thomas, 7, 44, 45, 46
electricity, 38, 39, 53, 54-55, 56, 63
electromagnetic waves, 52
endoscope, 43
engine, 33, 48
eyeglasses, 28

F G

Faraday, Michael, 54
film, 41, 50, 51, 60
fire, 8, 10, 14, 20, 21
flight, 58-59
flint, 10, 11, 18, 21
food canning, 6
food mixer, 54, 55
four-stroke engine, 48, 49
Franklin, Benjamin, 38
furnace, 14, 15, 25
Galilei, Galileo, 28
Galvani, Luigi, 38
Gassner, Carl, 39
gear wheels, 8, 9, 25
generators, 54
glass making, 6
Gutenberg, J., 26, 27

H

Hadley, John, 34
hair dryer, 55
Hall, Chester Moor, 28
Halladay Standard Windmill, 25
Harrington, Sir John, 54
Harvey, William, 43
heating element, 54
helicopter, 58
Henson, William, 58, 59
Hero of Alexandria, 32
Hertz, Heinrich, 52
Hoover, William, 55
horse power, 24
horseshoe, 15
hot-air balloon, 58
Huygens, Christian, 23

I J K

information technology, 62
ink, 18, 19
internal combustion engine, 48-49
iron, 14, 15, 16, 44, 46, 55
Judson, Whitcomb, 7
kettle, 54

L

Laennec, René, 43
lantern clock, 23
latitude, 34, 35
lead-acid accumulator, 39
Lenoir, Etienne, 48
lenses, 28, 29, 40, 41, 50
letterpress, 26
light, 28-29
lighthouse, 35
lighting, 7, 20-21, 54
lightning, 38
Lilienthal, Otto, 58
limelight, 50
Lister, Joseph, 43
lock, 7
locomotive, 32, 33
lodestone, 34
logarithms, 30
longtitude, 34, 35
loom, 36, 37
Loud, John H., 19
loudspeakers, 52
Lumière, Auguste and Louis, 50

M

Macarius, Joannes, 35
magic lanterns, 50
magnet, 44
magnetic tape, 46, 47
Marconi, Guglielmo, 52
matches, 7, 21
measurement, 16-17, 34
mechanical clocks, 22
medical instruments, 42-43
Meikle, Andrew, 25
merkhet, 22
metalworking, 14-15
microchip, 62-63
microelectronics, 46, 62-63
microphone, 52
microscope, 28, 29, 63
mirrors, 28, 29, 35
Montgolfier brothers, 58
Muybridge, Eadweard, 50, 51

N O

Napier, John, 31
navigation, 34-35
needles, 15, 42, 46, 47
Newcomen, Thomas, 32, 33
Niepce, Joseph
Nicéphore, 40
Nipkow, Paul, 56
octant, 34, 35
oil lamps, 20, 21
optics, 28-29, 40
ores, 14
Otto, Nikolaus, 48

P Q

paper, 7, 19
Papin, Denis, 55
papyrus, 18
Parkes, Alexander, 60
Pascal, Blaise, 31
pencil, 7, 18
pendulums, 22, 23
pens, 18, 19
phonograph, 46
photography, 40-41, 50
Planté, Gaston, 39
plastics, 60-61, 62
plow, 7
plumb line, 29
potter's wheel, 12
Poulsen, Valdemar, 47

power station, 54
Pratt, William, 31
Pravaz, Charles Gabriel, 42
printed circuit board, 63
printing, 18, 26-27
prism, 29, 51
quadrant, 29

R

radio, 52-53, 62
recording, 46-47
refraction, 28
refrigerators, 54
Roget, P. M., 50
rollers, 12, 13
Röntgen, Wilhelm, 56
rulers, 11, 16, 17

S

sails, 25, 33
sandglass, 23
satellite, 63
Savery, Thomas, 32
saws, 11, 42
scales, 16
Schulze, Johann Heinrich, 40
scissors, 6
sextant, 35
silicon chips, 62
smart card, 63
sound vibrations, 44, 45, 46
spacecraft, 62, 63
sphygmomanometer, 43
spinning, 36-37
steam, 32-33, 37, 55
steam engines, 12, 32-33, 48, 58, 59
Stephenson, George, 33
stethoscope, 42, 43
stone tools, 10-11
Stringfellow, John, 58, 59
stylus, 18, 31, 46, 47
Sundback, Gideon, 7
sundial, 22
surveying, 34, 35
Swan, Joseph, 7

T

Tainter, Charles, 47
Talbot, W. H. Fox, 40
tally sticks, 30
tape recording, 46, 47

telegraphy, 44, 52
telephone, 44-45, 46, 52
telescope, 28, 29
television, 53, 56-57, 62
terminal, 63
Thales, 38
theodolite, 34
thermometer, 43
tinder box, 21
toilet, 54
Tompion, Thomas, 23
tools, 8, 10-11
transmitters, 52, 53
treadmills, 24
Trevithick, Richard, 33
trigonometry, 34
T'sai Lun, 7
typecasting, 26-27

V

vacuum cleaners, 54, 55
vacuum tubes, 52, 53, 56
Van Leeuwenhoek
Antoni, 29
Verne, Jules, 59
Volta, Alessandro, 38
von Basch, Samuel, 43

W

watchmaking, 23
water power, 24, 25, 37
water pump, 32
Waterman, Edson, 19
Watt, James, 32, 33
wax cylinder, 47
weaving, 36-37
welding, 14, 15
wheels, 12-13, 24, 25, 31, 48
windmills, 12, 24, 25
Wollaston, W. H., 38
Wright brothers, 58, 59
writing implements, 18-19

X Y Z

X-rays, 52, 56
yardstick, 17
Zeiss, Carl, 41
zipper, 7

Acknowledgments

Dorling Kindersley would like to thank:
The following members of the staff of the Science Museum, London for help with the provision of objects for photography and checking the text: Marcus Austin, Peter Bailes, Brian Bowers, Roger Bridgman, Neil Brown, Jane Bywaters, Sue Cackett, Janet Carding, Ann Carter, Jon Darius, Eryl Davies, Sam Evans, Peter Fitzgerald, Jane Insley, Stephen Johnston, Ghislaine Lawrence, Peter Mann, Mick Marr, Kate Morris, Susan Mossman, Andrew Nahum, Cathy Needham, Francesca Riccini, Derek Robinson, Peter Stephens, Frazer Swift, Peter Tomlinson, John Underwood, Denys Vaughan, Tony Vincent, John Ward, Anthony Wilson, David Woodcock, Michael Wright.

Retouching Roy Flooks

Index Jane Parker

Picture credits
t = top, b = bottom, m = middle, l = left, r = right

Bridgeman Art Library: 11, 18bm, 19bl; /Russian Museum, Leningrad 21 mr, 22tr; /Giraudon/Musée des Beaux Arts, Vincennes 30bl, 50mr
E.T. Archive: 26tr
Mary Evans Picture Library: 10m, 12ml, 12tr, 14, 19mr, 19br, 20tr, 21mr, 23tr, 24tr, 25tr, 28br, 30bl, 31mr, 36bl, 39m, 40tl, 40mr, 41tm, 42bl, 42ml, 43tr, 43m, 45tr, 50br, 53mr, 54tl, 54bl, 55ml
Vivien Fifield: 32m, 48ml, 48m, 48bl
Michael Holford: 16ml, 18,lm, 18bl
Hulton-Deutsch: 41tr
National Motor Museum Beaulieu: 49tr
Ann Ronan Picture Library: 17tl, 29br, 29bm, 35br, 38tl, 38mr, 44tl, 44tr, 45tl, 56br

Science Photo Library: 63m, 63bl, 63bm
Syndication International: 12tl, 13m, 23bl, 26mr, 28tl, 28mr, 34tl, 35tl, 36tr, 46br, 50br, 52b, 52tr, 56tr, 58bm, 59br; / Bayerische Staatsbibliotek, Munich 24cl; /City of Bristol Museum and Art Gallery 53br; /British Museum 13tm, 24bl, 59tr; / Library of Congress 37tl; /Smithsonian Institution, Washington DC 58br

With the exception of the items listed above, and the objects on pages 8-9 and 61, all the photographs in this book are of objects in the collections of the Science Museum, London.